FIELD GUIDE FOR STREAM CLASSIFICATION

Prepared By:
Dave Rosgen and Lee Silvey
Wildland Hydrology

©1998 All rights reserved.
ISBN 0-9653289-1-0

Book and Cover Design: Colorline

FIELD GUIDE FOR STREAM CLASSIFICATION

FOREWORD

FIELD GUIDE FOR STREAM CLASSIFICATION

This field guide for stream classification is designed as a visual reference document to assist in the delineation of a variety of stream types. The delineation keys, photographs and illustrations that describe morphological characteristics are taken from Applied River Morphology, by Dave Rosgen and H. Lee Silvey. Our experience has shown that visual comparisons of rivers having similar morphology assists in the proper classification of rivers being studied. The presentation of frequency distributions for delineative criteria by stream type also provide a range of data for comparison with streams of similar character.

Verification of stream types involves field measurements of morphological criteria which provides an essential link with the visual guides. Field forms for classification involving dimensions, pattern, profile and channel materials are provided in a custom designed book titled, "The Reference Reach Field Book." The field book documents and provides a permanent record of river data. The field book also allows documentation of stream gaging stations for calibration of bankfull stage, hydraulic geometry plots and for construction of "regional curves." This field book is available through Wildland Hydrology Books, 1481 Stevens Lake Road, Pagosa Springs, Colorado, 81147.

The river data that is obtained from your field efforts will form a data base and provide a better understanding of river morphology for management applications. The correct classification of rivers will allow one to: 1) predict a river's behavior from its appearance; 2) develop empirical relations for individual stream types; 3) stratify and analyze companion inventory data; 4) extrapolate data from other rivers of similar stream type; and 5) communicate more effectively with others concerned with the river resource.

Field measurements and their attendant data will provide the foundation for the knowledge of river process, and hopefully, for those who have yet to venture forth to the field, the enlightenment necessary to avoid the common misconceptions often engendered in the office-bound darkness.

DAVE ROSGEN AND H. LEE SILVEY
Wildland Hydrology

FIELD GUIDE FOR STREAM CLASSIFICATION

TABLE OF CONTENTS

FIELD GUIDE FOR STREAM CLASSIFICATION

General Stream Classification Information
- Heirarchical River Inventories 1
- Broad Level (I) Stream Classification 2-3
- General Stream Type Descriptions (Level I) 4-5
- Valley Types 6-20
- Level II Classification Criteria (Field form sample) 21
- Primary Delineative Criteria 22
- Classification Key 23

The Bankfull Stage
- Locations to Measure Bankfull Stage 24-25
- Examples of Bankfull Stage 26-30

Delineative Criteria
- Entrenchment Ratio... 31
- Field Procedures for Determining Entrenchment 32
- Stream Channel Cross-Section 32
- Width/Depth Ratio... 34
- Representative Pebble Count Procedure... 35
- Channel Sinuosity Calculation 36
- Slope Measurement 37

Stream Type Photographs and Illustrations
- Morphological Descriptions and Examples of Stream Types 39
  - Stream Types A1-A6 40-63
  - Stream Types B1-B6 64-87
  - Stream Types C1-C6 88-111
  - Stream Types D3-D6 112-125
  - Stream Types DA-DA6 126-129
  - Stream Types E3-E6 130-145
  - Stream Types F1-F6 146-169
  - Stream Types G1-G6 170-193

FIELD GUIDE FOR STREAM CLASSIFICATION

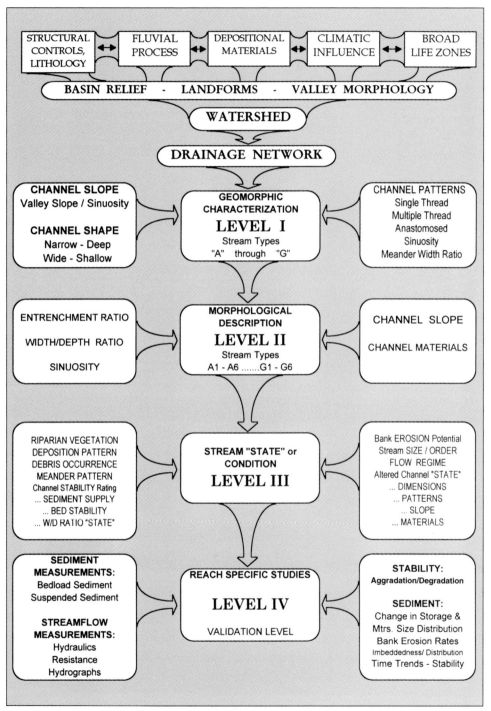

FIGURE 1. Hierarchical river inventory levels

FIELD GUIDE FOR STREAM CLASSIFICATION

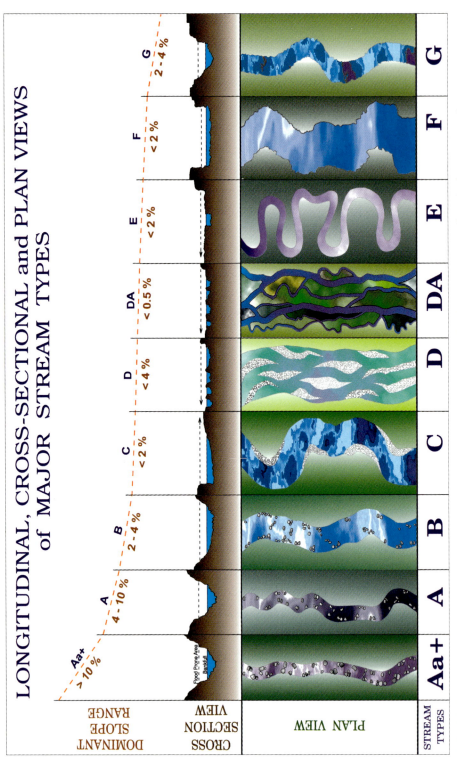

FIGURE 2. Broad level stream classification delineation showing longitudinal, cross-sectional and plan views of major stream types. (from Rosgen, 1994)

FIELD GUIDE FOR STREAM CLASSIFICATION

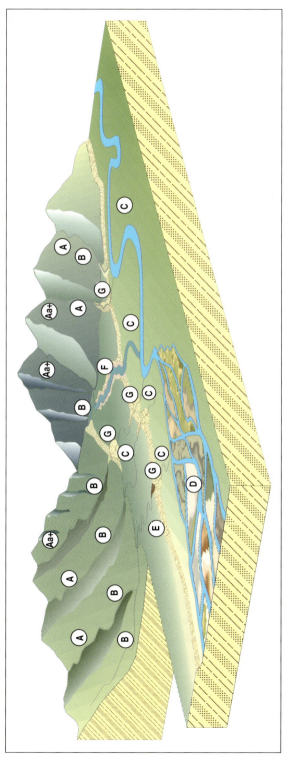

FIGURE 3. Example of broad level delineation of stream types at Level I.

FIELD GUIDE FOR STREAM CLASSIFICATION

Stream Type	General Description	Entrenchment Ratio	W/D Ratio	Sinuosity	Slope %	Landform/Soils/Features
Aa+	Very steep, deeply entrenched, debris transport, torrent streams.	<1.4	<12	1.0 to 1.1	>10	Very high relief. Erosional, bedrock or depositional features; debris flow potential. Deeply entrenched streams. Vertical steps with deep scour pools; waterfalls.
A	Steep, entrenched, cascading, step/pool streams. High energy/debris transport associated with depositional soils. Very stable if bedrock or boulder dominated channel.	<1.4	<12	1.0 to 1.2	4 to 10	High relief. Erosional or depositional and bedrock forms. Entrenched and confined streams with cascading reaches. Frequently spaced, deep pools in associated step/pool bed morphology.
B	Moderately entrenched, moderate gradient, riffle dominated channel, with infrequently spaced pools. Very stable plan and profile. Stable banks.	1.4 to 2.2	>12	>1.2	2 to 4	Moderate relief, colluvial deposition, and/or structural. Moderate entrenchment and W/D ratio. Narrow, gently sloping valleys. Rapids predominate w/scour pools.
C	Low gradient, meandering, point-bar, riffle/pool, alluvial channels with broad, well defined floodplains.	>2.2	>12	>1.2	<2	Broad valleys w/terraces, in association with floodplains, alluvial soils. Slightly entrenched with well-defined meandering channels. Riffle/pool bed morphology.
D	Braided channel with longitudinal and transverse bars. Very wide channel with eroding banks.	n/a	>40	n/a	<4	Broad valleys with alluvium, steeper fans. Glacial debris and depositional features. Active lateral adjustment, w/abundance of sediment supply. Convergence/divergence bed features, aggradational processes, high bedload and bank erosion.

TABLE 1. General stream type descriptions and delineative criteria for broad-level classification. (Level I)

FIELD GUIDE FOR STREAM CLASSIFICATION

Stream Type	General Description	Entrenchment Ratio	W/D Ratio	Sinuosity	Slope %	Landform/Soils/Features
DA	Anastomosing (multiple channels) narrow and deep with extensive, well vegetated floodplains and associated wetlands. Very gentle relief with highly variable sinuosities and width/depth ratios. Very stable streambanks.	>2.2	Highly variable	Highly variable	<.5	Broad, low-gradient valleys with fine alluvium and/or lacustrine soils. Anastomosed (multiple channel) geologic control creating fine deposition w/well-vegetated bars that are laterally stable with broad wetland floodplains. Very low bedload, high wash load sediment.
E	Low gradient, meandering riffle/pool stream with low width/depth ratio and little deposition. Very efficient and stable. High meander width ratio.	>2.2	<12	>1.5	<2	Broad valley/meadows. Alluvial materials with floodplains. Highly sinuous with stable, well-vegetated banks. Riffle/pool morphology with very low width/depth ratios.
F	Entrenched meandering riffle/pool channel on low gradients with high width/depth ratio.	<1.4	>12	>1.2	<2	Entrenched in highly weathered material. Gentle gradients, with a high width/depth ratio. Meandering, laterally unstable with high bank erosion rates. Riffle/pool morphology.
G	Entrenched "gully" step/pool and low width/depth ratio on moderate gradients.	<1.4	<12	>1.2	2 to 4	Gullies, step/pool morphology w/moderate slopes and low width/depth ratio. Narrow valleys, or deeply incised in alluvial or colluvial materials, i.e., fans or deltas. Unstable, with grade control problems and high bank erosion rates.

TABLE 1. (Cont.) General stream type descriptions and delineative criteria for broad-level classification. (Level I)

FIELD GUIDE FOR STREAM CLASSIFICATION

FIGURE 5-a. *Valley Type II*, moderately steep, gentle sloping side slopes often in colluvial valleys (B stream types).

FIGURE 4a. *Valley Type I*, "V" notched canyons, rejuvenated sideslopes (A and G stream types).

FIELD GUIDE FOR STREAM CLASSIFICATION

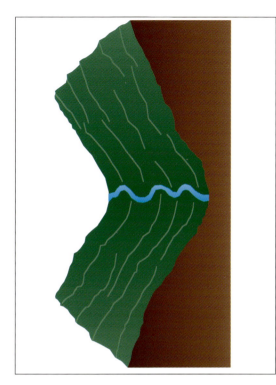

FIGURE 5b. *Valley Type II*, moderately steep, gentle sloping side slopes often in colluvial valleys.

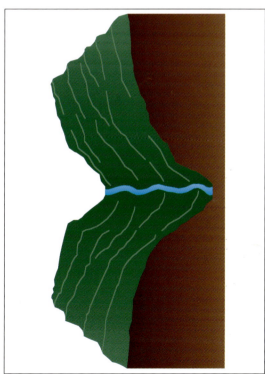

FIGURE 4b. *Valley Type I*, "V" notched canyons, rejuvenated sideslopes.

FIELD GUIDE FOR STREAM CLASSIFICATION

FIGURE 7a. *Valley Type IV*, gentle gradient canyons, gorges and confined alluvial valleys (F or C stream types).

FIGURE 6a. *Valley Type III*, alluvial fans and debris cones (A, G, D and B stream types).

8

FIELD GUIDE FOR STREAM CLASSIFICATION

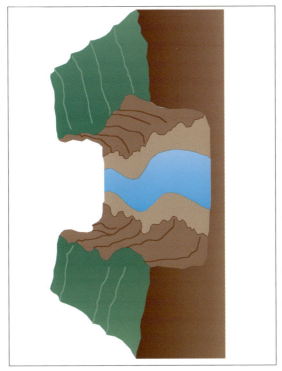

FIGURE 7b. *Valley Type IV*, gentle gradient canyons, gorges and confined alluvial valleys.

FIGURE 6b. *Valley Type III*, alluvial fans and debris cones.

FIELD GUIDE FOR STREAM CLASSIFICATION

FIGURE 9a. *Valley Type VI*, moderately steep, fault controlled valleys (B, G and C stream types).

FIGURE 8a. *Valley Type V*, moderately steep valley slopes, "U" shaped glacial trough valleys (D and C stream types).

FIELD GUIDE FOR STREAM CLASSIFICATION

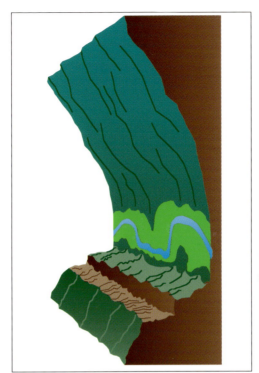

FIGURE 9b. *Valley Type VI*, moderately steep, fault controlled valleys.

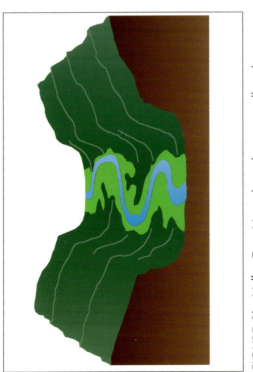

FIGURE 8b. *Valley Type V*, moderately steep valley slopes, "U" shaped glacial trough valleys.

11

FIELD GUIDE FOR STREAM CLASSIFICATION

FIGURE 11a. *Valley Type VIII*, wide, gentle valley slope with a well developed floodplain adjacent to river terraces.

FIGURE 10a. *Valley Type VII*, steep, highly dissected fluvial slopes (A and G stream types).

FIELD GUIDE FOR STREAM CLASSIFICATION

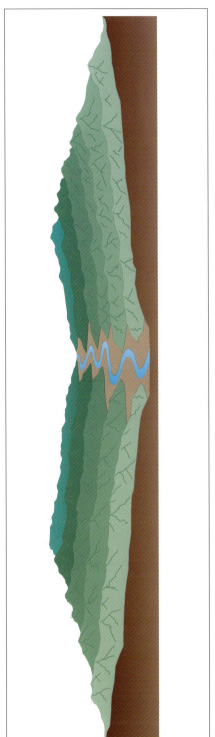

FIGURE 10b. *Valley Type VII*, steep, highly dissected fluvial slopes.

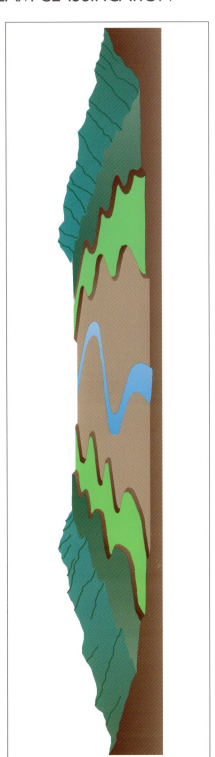

FIGURE 11b. *Valley Type VIII*, wide, gentle valley slope with well developed floodplain adjacent to river terraces.

FIELD GUIDE FOR STREAM CLASSIFICATION

FIGURE 13a. *Valley Type X*, very broad and gentle slopes, associated with extensive floodplains - Great Plains, semi-desert and desert provinces; coastal plains and tundra; Lacustrine valleys.

FIGURE 12a. *Valley Type IX*, broad, moderate to gentle slopes, associated with glacial outwash and/or eolian sand dunes (Predominately D and some C stream types).

14

FIELD GUIDE FOR STREAM CLASSIFICATION

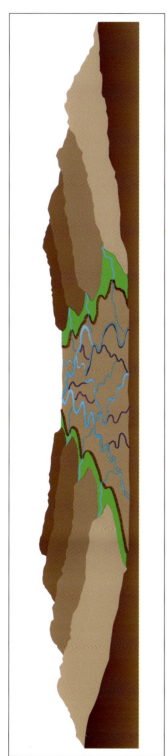

FIGURE 12b. *Valley Type IX*, broad, moderate to gentle slopes, associated with glacial outwash and/or eolian sand dunes.

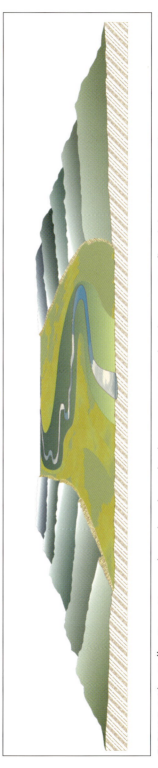

FIGURE 13b. *Valley Type X*, very broad and gentle slopes, associated with extensive floodplains - Great Plains, semi-desert and desert provinces; coastal plains and tundra; Lacustrine valleys.

FIELD GUIDE FOR STREAM CLASSIFICATION

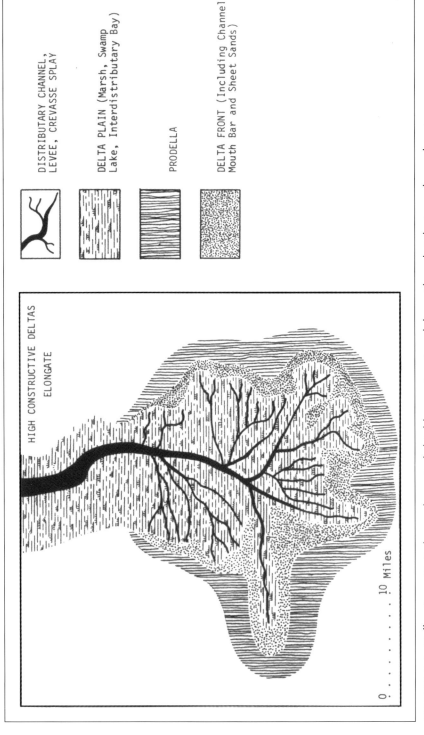

FIGURE 14. *Valley Type XI, Deltas - elongated, highly constructive delta with a distributary channel system.* (Fisher et al., 1969)

FIELD GUIDE FOR STREAM CLASSIFICATION

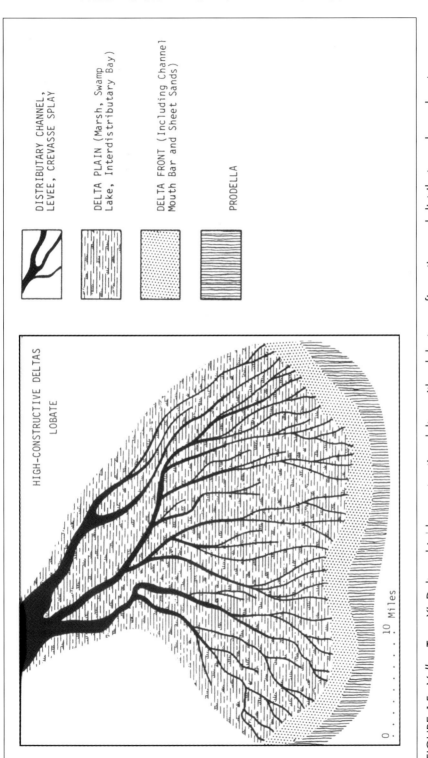

FIGURE 15. *Valley Type XI, Deltas* - highly constructive deltas with a lobate configuration and distributary channel system. (Fisher et al., 1969)

FIELD GUIDE FOR STREAM CLASSIFICATION

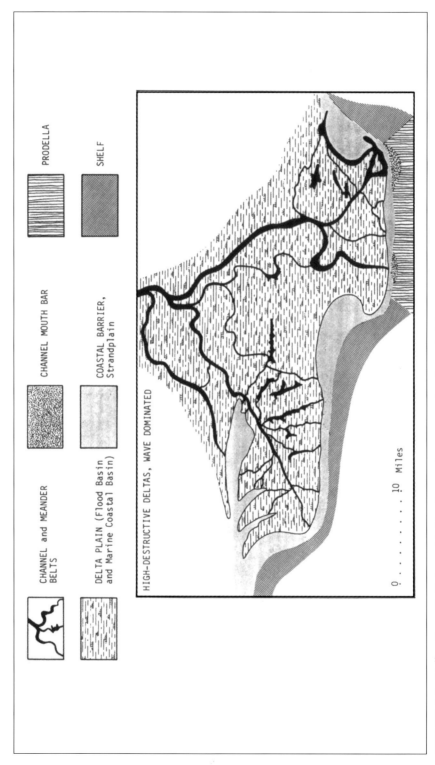

FIGURE 16. *Valley Type XI, Deltas - highly destructive, wave dominated delta. (Fisher, et al., 1969)*

FIELD GUIDE FOR STREAM CLASSIFICATION

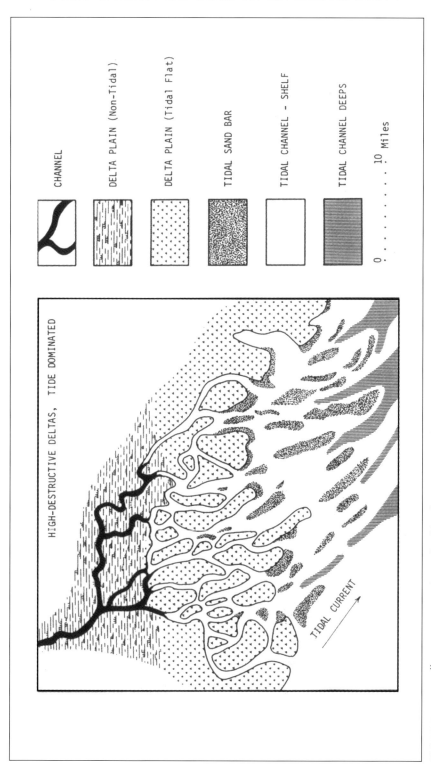

FIGURE 17. Valley Type XI, Deltas - highly destructive, tide dominated delta. *(Fisher et al., 1969)*

FIELD GUIDE FOR STREAM CLASSIFICATION

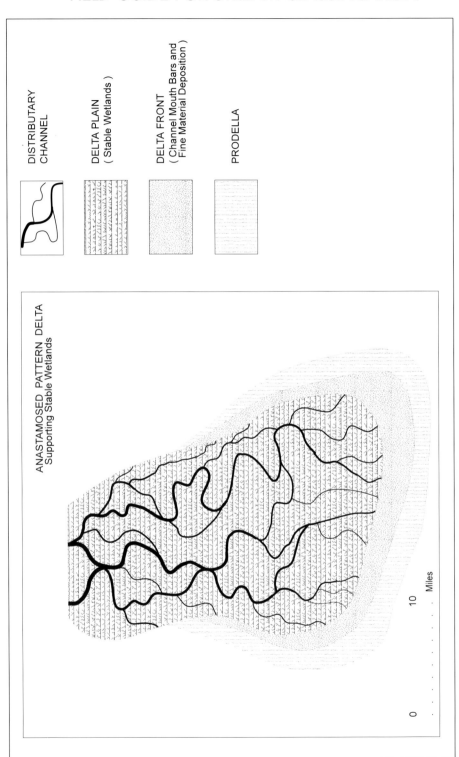

FIGURE 18. *Valley Type XI*, Deltas - anastomosed river delta pattern with supporting stable wetlands and channels.

FIELD GUIDE FOR STREAM CLASSIFICATION

Stream Channel Classification (Level II) ....

Stream NAME: _____
Basin NAME: _____ Drainage AREA: _____ Ac. _____ SqMi.
Location: _____
Twp: _____ Rge: _____ Sec: _____ Qtr: _____ Lat. _____ Long. _____
Observers: _____ Date: _____

Bankfull WIDTH (W_{bkf}) _____ Ft.
WIDTH of the stream channel, at bankfull stage elevation, in a riffle section.

Mean DEPTH (d_{bkf}) _____ Ft.
Mean DEPTH of the stream channel cross-section, at bankfull stage elevation, in a riffle section. (d_{bkf} = A / W_{bkf})

Bnkfl. X-Section AREA (A_{bkf}) _____ Sq.Ft.
AREA of the stream channel cross-section, at bankfull stage elevation, in a riffle section.

Width / Depth RATIO (W_{bkf} / d_{bkf}) _____
Bankfull WIDTH divided by bankfull mean DEPTH, in a riffle section.

Maximum DEPTH (d_{mbkf}) _____ Ft.
Maximum depth of the bankfull channel cross-section, or distance between the bankfull stage and thalweg elevations, in a riffle section.

WIDTH of Flood-Prone Area (W_{fpa}) _____ Ft.
Twice maximum DEPTH, or (2 x d_{mbkf}) = the stage/elevation at which flood-prone area WIDTH is determined. (riffle section)

Entrenchment Ratio (ER) _____
The ratio of flood-prone area WIDTH divided by bankfull channel WIDTH. (W_{fpa} / W_{bkf}) (riffle section)

Channel Materials *(Particle Size Index)* D50 _____ mm.
The D50 particle size index represents the **median** diameter of channel materials, as sampled from the channel surface, between the bankfull stage and thalweg elevations.

Water Surface SLOPE (S) _____ Ft./Ft.
Channel slope = "rise" over "run" for a reach approximately 20 - 30 bankfull channel widths in length, with the "riffle to riffle" water surface slope representing the gradient at bankfull stage.

Channel SINUOSITY (K) _____
Sinuosity is an index of channel pattern, determined from a ratio of stream length divided by valley length (SL/ VL); or estimated from a ratio of valley slope divided by channel slope (VS/S).

Stream Type _____

For reference, note: StreamType Chart & Classification Key

TABLE 2. Level II classificaton criteria, (field form)

FIELD GUIDE FOR STREAM CLASSIFICATION

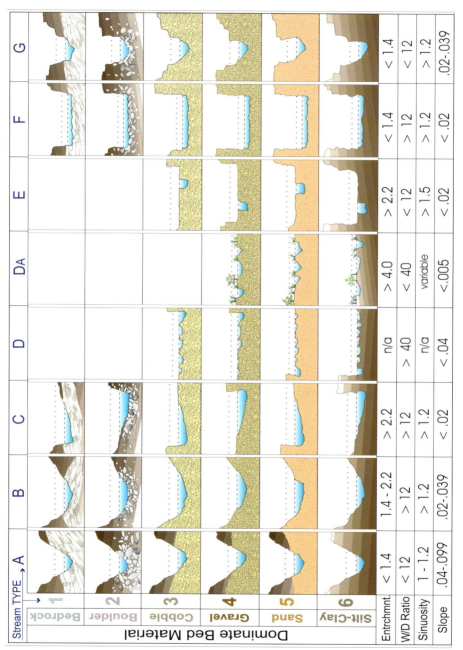

FIGURE 19. Primary delineative criteria for the major stream types.

FIELD GUIDE FOR STREAM CLASSIFICATION

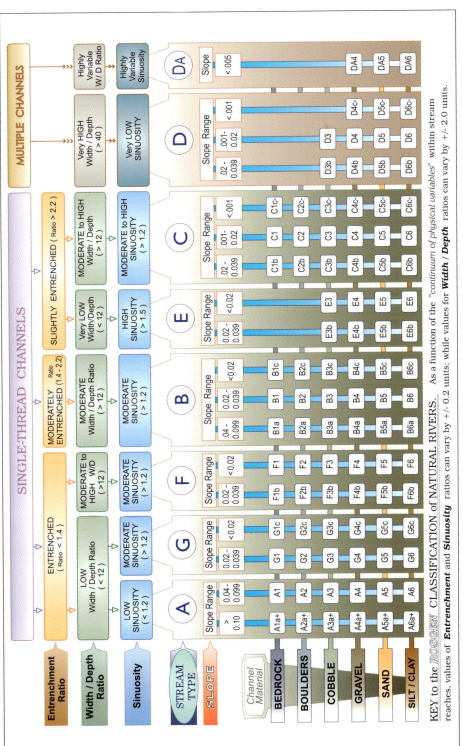

FIGURE 20. Classification key for natural rivers.

FIELD GUIDE FOR STREAM CLASSIFICATION

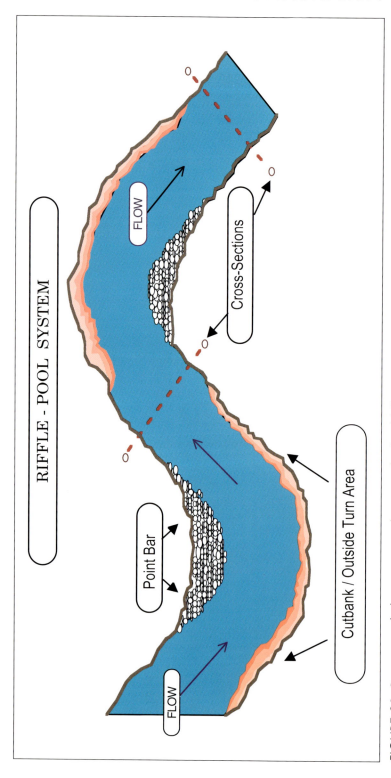

FIGURE 21. Recommended cross-section locations for bankfull stage measurements in "riffle/pool" systems.

FIELD GUIDE FOR STREAM CLASSIFICATION

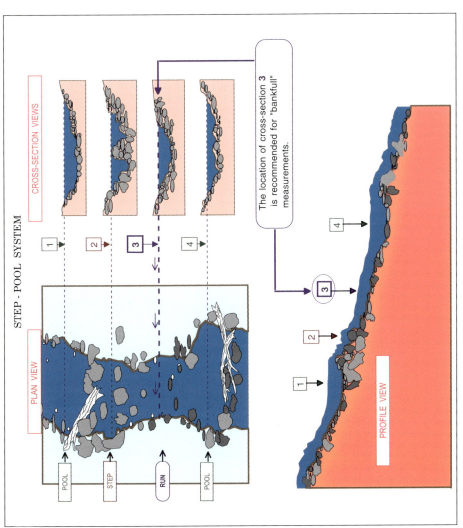

FIGURE 22. Recommended location for measurement of bankfull stage in "riffle/pool" systems.

FIELD GUIDE FOR STREAM CLASSIFICATION

a) Conifers as indicators of bankfull stage, B4 Stream type.

b) Top of point bar corresponding to woody species indicating the bankfull stage. Floodplain on right. C5 Stream type.

FIGURE 23a-b. Photographs of various stream types depicting bankfull stage and corresponding indicators.

FIELD GUIDE FOR STREAM CLASSIFICATION

c) Highest central bars coincide with bank on right, D3 stream type - Blanco River.

d) Base of alders and change in bank slope in Stream type B3 - Leche Creek.

FIGURE 23c-d. Photographs of various stream types depicting bankfull stage and corresponding indicators.

FIELD GUIDE FOR STREAM CLASSIFICATION

e) C5 Stream type showing top of point bars and riparian plant indicators of the bankfull stage.

f) Top of bank and "bankfull" at same level, E3 Stream type - O'Neill Creek.

FIGURE 23e-f. Photographs of various stream types depicting bankfull stage and corresponding indicators.

FIELD GUIDE FOR STREAM CLASSIFICATION

a) Note staining on rocks which correspond to brush/rock interface at the bankfull stage - B3 Stream Type, Lake Creek, Alaska.

b) Note top of point bar and rock "line" on opposite right bank delineating the bankfull stage - F4 Stream Type, Duchesne River, Utah. (*Photo by J. Winston*)

FIGURE 24a-b. Photographs of various stream types depicting bankfull stage and corresponding indicators.

FIELD GUIDE FOR STREAM CLASSIFICATION

c) Note top of point bar indicating bankfull stage in entrenched G4 Stream type - Rito Blanco, Colorado.

d) Top of Point bar and willows indicating bankfull stage in C4 Stream Type - Upper Willow Creek, Colorado.

FIGURE 24c-d. Photographs of various stream types depicting bankfull stage and corresponding indicators.

FIELD GUIDE FOR STREAM CLASSIFICATION

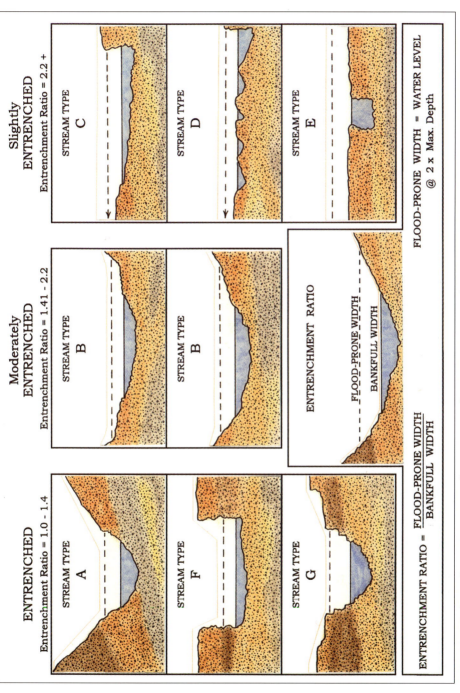

FIGURE 25. Representative entrenchment ratios for cross-sections of various stream types.

FIELD GUIDE FOR STREAM CLASSIFICATION

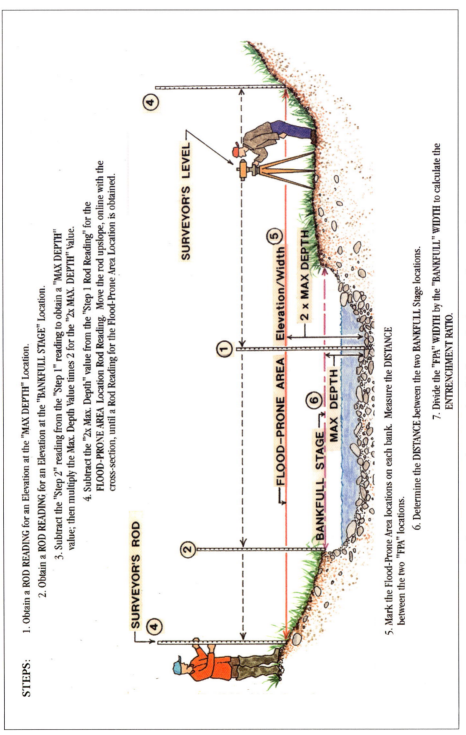

FIGURE 26. Determining a Flood-Prone Area elevation/width for calculation of Entrenchment Ratio.

FIELD GUIDE FOR STREAM CLASSIFICATION

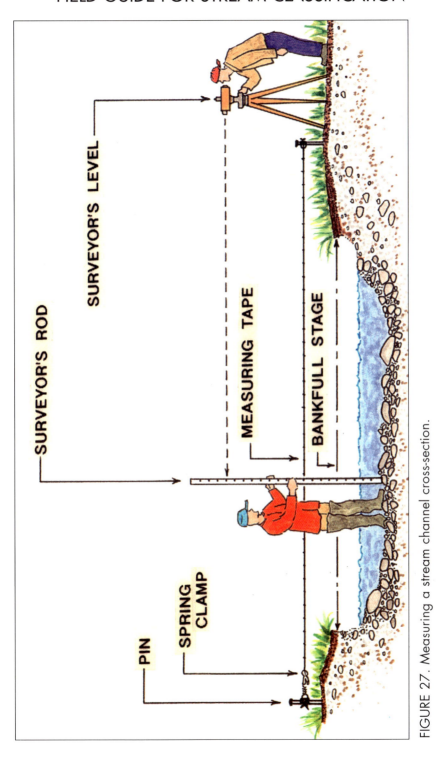

FIGURE 27. Measuring a stream channel cross-section.

FIELD GUIDE FOR STREAM CLASSIFICATION

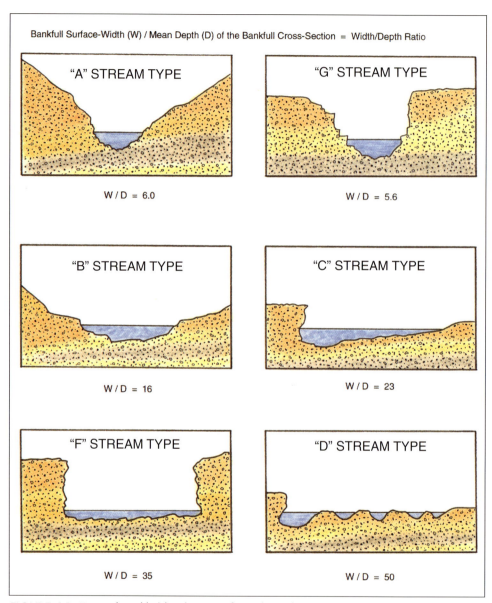

FIGURE 28. Typical width/depth ratios for selected stream types.

FIELD GUIDE FOR STREAM CLASSIFICATION

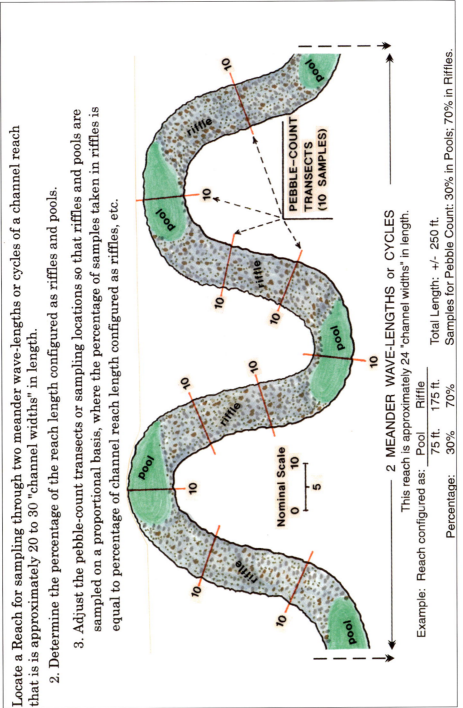

FIGURE 29. Representative Pebble-Count procedure.

35

FIELD GUIDE FOR STREAM CLASSIFICATION

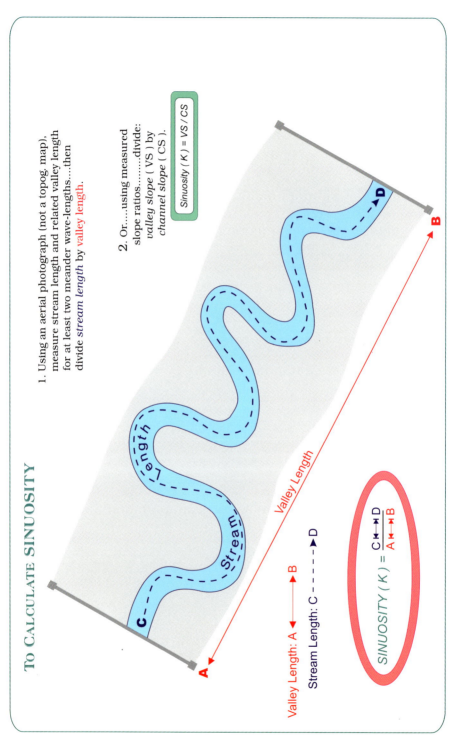

FIGURE 30. Channel Sinuosity Calculations

FIELD GUIDE FOR STREAM CLASSIFICATION

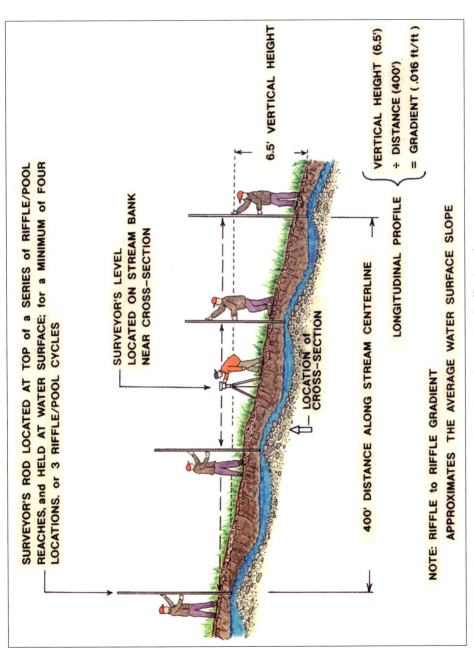

FIGURE 31. Measuring stream gradient through a typical riffle/pool sequence.

MORPHOLOGICAL DESCRIPTIONS AND EXAMPLES OF STREAM TYPES

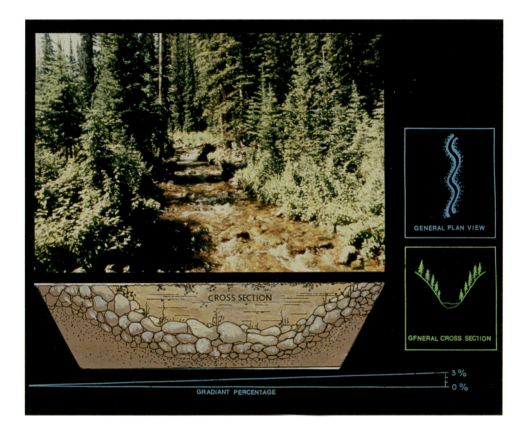

MORPHOLOGICAL DESCRIPTION AND EXAMPLES OF STREAM TYPES

A1 Stream Type

The A1 stream type is a steep, entrenched and confined channel in bedrock, that is associated with faults, scarps, folds, joints and other structurally controlled drainage ways. The A1 stream type is typically found in a valley where stream slopes can range from 4-10%. The stream bed is seen as a cascading, step/pool morphology with irregularly spaced drops and deep scour pools. Channel gradients greater than 10% are categorized as A1a+ and may produce streamflow described as torrents or with torrential flows, waterfalls, and bed forms described as "chutes," or a series of vertical drops. Pool spacing is highly irregular and is controlled by bedrock and large, woody organic debris.

As gradient increases, the steps become more closely spaced. The width/depth ratios are characteristically low; however, some observations indicate width/depth ratios greater than 12, due to bedrock sills. The A1 stream type is relatively straight with a sinuosity less than 1.2 Channel materials are principally bedrock although boulders and, in smaller amounts, cobbles and gravels may be included.. The dimension, pattern, and profile of this stream type is little influenced by the frequent flows and sediment regime. The A1 stream type is a high energy, low sediment supply stream system.

THE MORPHOLOGICAL DESCRIPTION

DELINEATIVE CRITERIA (A1)

Landform/soils: Bedrock controlled, steep slopes and channel. Glacial scoured slopes, faults, folds and joints. Soils predominantly bedrock and coarse colluvium.

Channel materials: Bedrock with lesser amounts of boulders, cobble and gravel.

Slope Range: .04 - .10 (A1a+ > .10) **Entrenchment Ratio:** < 1.4

Width/depth Ratio: < 12 **Sinuosity:** < 1.2

THE MORPHOLOGICAL DESCRIPTION

A1. Washington

A1. Colorado

A1. Colorado

A1. Wyoming

THE MORPHOLOGICAL DESCRIPTION

A1. Idaho

A1. Colorado

A1. Wyoming

A1. Wyoming

MORPHOLOGICAL DESCRIPTION AND EXAMPLES OF STREAM TYPES

A2 Stream Type

Colorado

The A2 stream type is a steep, deeply entrenched and confined stream channel associated with faults, scarps, folds, joints, and other structurally controlled drainageways. Land forms supporting A2 stream types include canyons, steep side slopes, talus fields, glacial moraines, lag deposits, and coarse colluvial deposition. The A2 stream type is normally situated in valley types I and II. Slope ranges are 4-10 %, producing channels that exhibit step/pool bed features. However, the A2 stream type also occurs on slopes greater than 10 %, (A2a+), which promote cascades and "chutes".

The sinuosity is low (<1.2) as is the width/depth ratio (<12). Width/depth ratios greater than 12 can occur where larger boulders contribute to channel bed and bank stability. Stream types are incised in predominantly boulder-sized channel material with lesser amounts of cobble and gravel materials present. The A2 stream type is a high energy and low sediment supply stream type, with corresponding low bedload transport rates. The channel bed and stream banks are normally stable and contribute little to sediment supply.

THE MORPHOLOGICAL DESCRIPTION

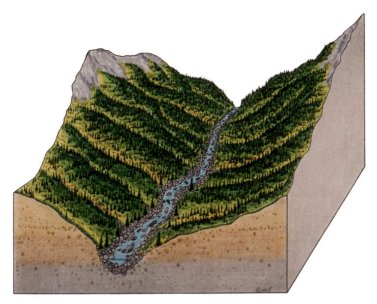

DELINEATIVE CRITERIA (A2)

Landform/soils: Steep, structural controlled slopes with colluvial deposition in narrow and confined valleys.

Channel materials: Predominantly boulders, with lesser amounts of cobble, gravel and sand. Some bedrock sporadically spaced.

Slope Range: .04 - .10 (A2a+ > .10)   **Entrenchment Ratio:** < 1.4

Width/depth Ratio: < 12   **Sinuosity:** < 1.2

THE MORPHOLOGICAL DESCRIPTION

STREAM TYPE A2

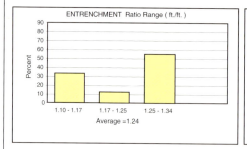

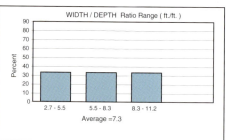

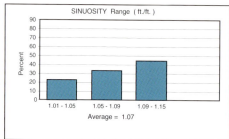

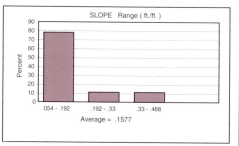

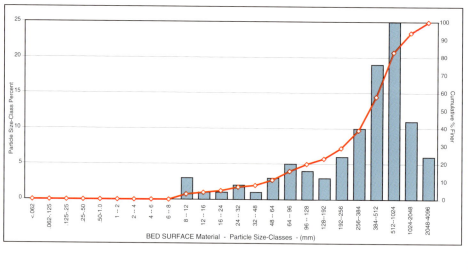

THE MORPHOLOGICAL DESCRIPTION

A2. Colorado

A2. Colorado

A2. Colorado

A2. Colorado

MORPHOLOGICAL DESCRIPTION AND EXAMPLES OF STREAM TYPES

A3 Stream Type

Colorado

The A3 stream type is a steep, deeply entrenched, and confined channel that is incised in coarse depositional soils. Landforms in which A3 types may be found are noted as glacial moraines or tills, alluvial fans, lag deposits, slide debris, and other coarse textured alluvial and/or colluvial deposition. The A3 stream type can also be situated as lateral tributary streams intersecting the valley train, and deeply incised in glacial and Holocene terraces adjacent to the river. The A3 stream type is observed in valley types I, III, and VII. The width depth ratios are low (< 12), with low sinuosity (< 1.2). The channel materials are typically unconsolidated, heterogenous, noncohesive materials, dominated by cobbles but also containing some small boulders, gravel, and sand. The A3 stream types develop both high energy (high stream power and shear stress values) and a high sediment supply, with corresponding very high bedload sediment transport rates. The A3 streams are generally unstable, with very steep, rejuvenated channel banks that contribute large quantities of sediment. The A3 bedform occurs as a step/pool, cascading channel which often stores large amounts of sediment in the pools associated with debris dams. The A3a+ stream types (slopes > 10%) are generally found in landforms associated with slump/earthflow and debris torrent/debris avalanche erosional processes. Characteristic stream bank erosional processes for the A3a+ stream type are fluvial entrainment, collapse (mass wasting), dry ravel, freeze/thaw, and debris flow scour.

THE MORPHOLOGICAL DESCRIPTION

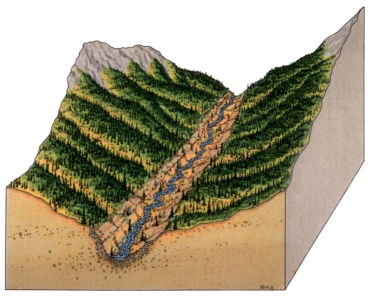

DELINEATIVE CRITERIA (A3)

Landform/soils: Steep, narrow depositional slopes typical of glacial moraines and debris slides associated with unconsolidated, heterogeneous and non-cohesive materials.

Channel materials: Predominantly cobble with a mixture of boulders, gravel and sand.

Slope Range: .04 - .10 (A3a+ > .10)   **Entrenchment Ratio:** < 1.4

Width/depth Ratio: < 12   **Sinuosity:** < 1.2

THE MORPHOLOGICAL DESCRIPTION

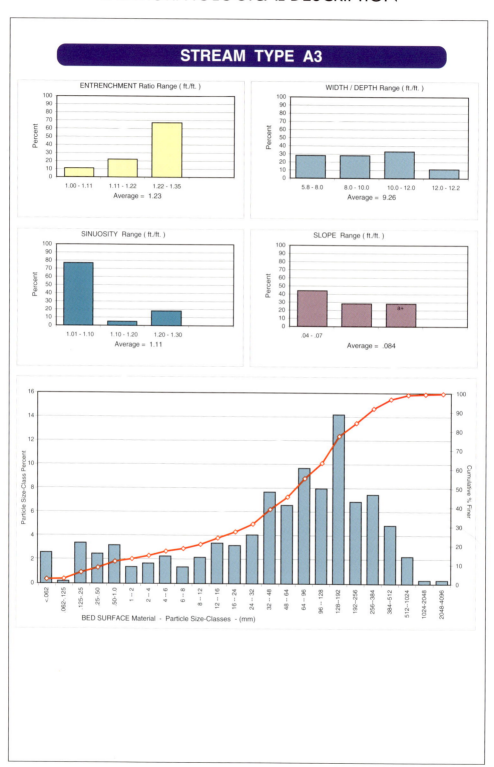

THE MORPHOLOGICAL DESCRIPTION

A3a+ - Colorado

A3 - Colorado

A3 - Colorado

51

MORPHOLOGICAL DESCRIPTION AND EXAMPLES OF STREAM TYPES

A4 Stream Type

The A4 stream type is a steep, deeply entrenched and confined, channel that is incised in coarse depositional materials. Landforms in which the A4 types develop are primarily glacial moraines or glacial tills, alluvial fans, lag deposits, landslide debris, and other coarse textured alluvial and/or colluvial deposition. The A4 types can also be situated as lateral tributary streams intersecting the valley train and deeply incised in glacial and Holocene river terraces. The A4 stream type appears in valley types I, III, and VII, in which some of the soils are residual and associated with highly weathered rock, such as grussic granite. The width depth ratios are low (< 12), with low sinuosity, (< 1.2). The A4 channel materials are typically unconsolidated, heterogenous, noncohesive materials, dominated by gravel, but also containing small amounts of boulders, cobble, and sand. The A4 channel bed features may be described as a step/pool or cascading channel, that is often influenced by the occurrence of organic woody debris that form debris dams, behind which are stored significant amounts of sediment in the pools. The A4 stream types typically have a high sediment supply which is combined with high energy streamflow to produce very high bedload sediment transport rates. The A4 stream types are generally unstable, with very steep, rejuvenated banks that contribute large quantities of sediment. Characteristic stream bank erosional processes for the A4 type are fluvial entrainment, bank collapse, dry ravel, freeze/thaw and lateral scour from debris flows. The A4a+ stream types (slopes > 10%) are usually located in slump/earthflow land forms and are often associated with debris avalanches and debris torrent erosional processes.

THE MORPHOLOGICAL DESCRIPTION

DELINEATIVE CRITERIA (A4)

Landform/soils: Steep, confined, depositional slopes in glacial and landslide debris. Soils heterogeneous mixture of unconsolidated colluvial and alluvial deposition.

Channel materials: Predominantly gravel with lesser amounts of boulders, cobble, and sand.

Slope Range: .04 - .10 (A4a+ > .10)   **Entrenchment Ratio:** < 1.4

Width/depth Ratio: < 12   **Sinuosity:** < 1.2

THE MORPHOLOGICAL DESCRIPTION

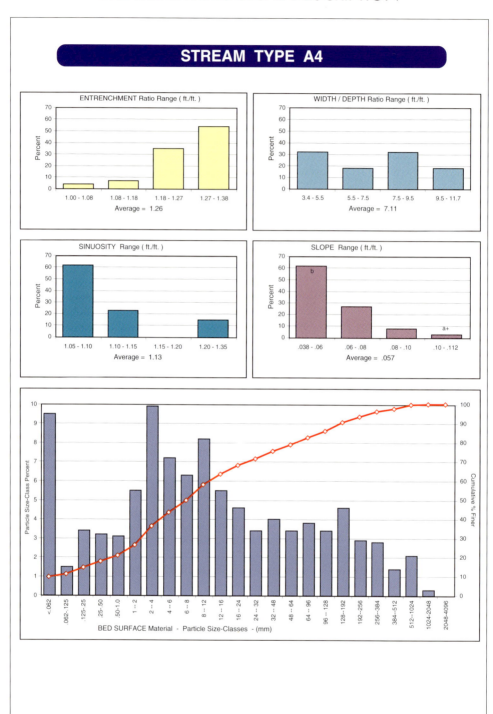

THE MORPHOLOGICAL DESCRIPTION

A4 - Colorado

A4 – Colorado

A4a+ - Colorado

A4a+ - California

MORPHOLOGICAL DESCRIPTION AND EXAMPLES OF STREAM TYPES

A5 Stream Type

Montana

The A5 stream types are steep, entrenched and confined channels, incised in predominantly sandy materials that are frequently intermixed with gravels. The A5 types are situated in both depositional landforms and residual soil areas that have developed from grussic granite and highly weathered sedimentary rocks. Glacio-fluvial, glacio-lacustrine, deltaic, eolian, and interbedded sedimentary deposits, as well as sorted sandy alluvium and unsorted sandy colluvium support the A5 stream type. Valley types that contain A5 stream types are I, III, and VII. The A5 channel has a low width/depth ratio, low sinuosity, and may be situated as a lateral channel to the valley train, or exist as an actively degrading main-stem channel in steep terrain. The channel bed and banks are unstable and very sensitive to induced changes in streamflow regime or in sediment supply. Bedload transport rates are very high, with streambank rejuvenation common. Bank erosion occurs through fluvial erosion, slump/earthflow, surface erosion, dry ravel, freeze/thaw and collapse. The A5a+ types (steeper than 10%) are often associated with mass-wasting erosional processes of debris avalanche and debris torrents.

THE MORPHOLOGICAL DESCRIPTION

DELINEATIVE CRITERIA (A5)

Landform/soils: Steep slopes in depositional landforms associated with erosional debris or residual soils derived from grussic granite and/or sandstones.

Channel materials: Predominantly sand with lesser amounts of gravel and silt/clay.

Slope Range: .04 - .10 (A5a+ > .10) **Entrenchment Ratio:** < 1.4

Width/depth Ratio: < 12 **Sinuosity:** < 1.2

THE MORPHOLOGICAL DESCRIPTION

A5a+ - Colorado

A5 - Wyoming

A5 - Nevada

A5a+ - Idaho

THE MORPHOLOGICAL DESCRIPTION

A5 - Colorado

A5 - Wyoming

A5 - Oregon

MORPHOLOGICAL DESCRIPTION AND EXAMPLES OF STREAM TYPES

A6 Stream Type

Colorado

The A6 stream type is a steep, entrenched and confined channel, incised in cohesive soils. Landforms and soils are observed as highly weathered shales and deposition from lacustrine, glacio-lacustrine, delta deposits, and fine grained alluvial deposits. The A6 stream type is found in valley types I, II, and VII. The channel width/depth ratio is the lowest of the A stream types due to the cohesive nature of the stream bank materials. The A6 stream type has a low sinuosity (< 1.2). Mass-wasting processes characteristic of the A6a channel banks (>10%) are creep, glide, slump/earthflow, and debris slides. The A6a+ stream types may also produce intense debris torrents. In contrast to the high rates of fluvial entrainment due to detachment of bed and bank materials of the A3, A4, and A5 stream types, the bank erosion of A6 streams is often due to liquefaction/saturation/collapse, and, negative pore-water pressure. The A6 stream type is normally associated with a step/pool profile and for the A6a+ as a cascade channel. Unlike the A3, A4, and A5 stream types, the A6 stream types do not develop a high bedload sediment yield. The A6 stream types are characteristic of a high contribution of wash load to the annual sediment yield.

THE MORPHOLOGICAL DESCRIPTION

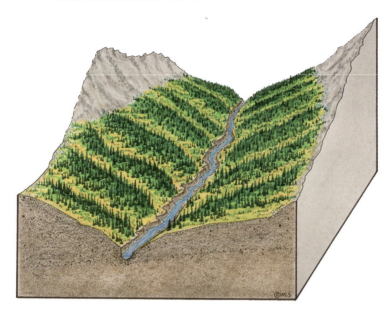

DELINEATIVE CRITERIA (A6)

Landform/soils: Steep slopes or channels dissected in fine alluvial silts and/or clays, or in residual soils derived from siltstone and shales.

Channel materials: Cohesive silt/clay with some sands.

Slope Range: .04 - .10 (A6a+ > .10)

Width/depth Ratio: < 12

Entrenchment Ratio: < 1.4

Sinuosity: < 1.2

THE MORPHOLOGICAL DESCRIPTION

A6a+ - California

A6 - Colorado

A6 - California

THE MORPHOLOGICAL DESCRIPTION

A6. - Utah

A6a+. Colorado

A6a+. California

MORPHOLOGICAL DESCRIPTION AND EXAMPLES OF STREAM TYPES

B1 Stream Type

The B1 stream type is a moderately entrenched channel associated with bedrock or bedrock controlled drainage ways, faults, folds, and joints, typically seen in valley types II and VI. While channel slopes normally range from 2 to 4%, the B1 type can occur on steeper slopes and still maintain a characteristically moderate (> 12) width/depth ratio. B1 stream types may also have channel slopes less than 2% and are categorized as B1c, to indicate that their morphology (e.g., width/depth ratio, confinement and sinuosity) is still characteristic of the B1 types despite the gentler gradient. Channel materials are dominated by bedrock but can also include boulders, cobble, and sand. The stream banks normally contain finer grained materials than the bed and are typically seen as an unconsolidated mixture of boulders, cobble, and sand. While the channel adjustment of the B1 stream type is primarily the result of lateral extension processes, channel sinuosity is not high (1.1-1.5), but is somewhat greater than the A1 stream type. The B1 stream types are considered as stable channel systems and contribute small amounts of sediment from their beds and banks. The B1 stream type channels are dominated by bed features that produce extensive rapids, with infrequent scour holes for pools. The sequence of the pool-to-pool spacing is irregular and infrequent due to the nature of the resident bedrock materials.

THE MORPHOLOGICAL DESCRIPTION

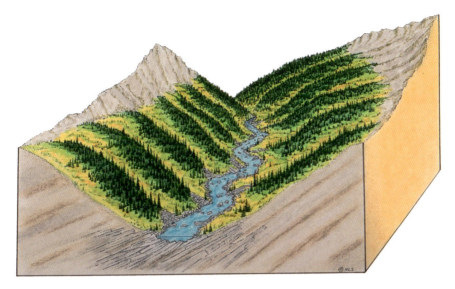

DELINEATIVE CRITERIA (B1)

Landform/soils: Structural controlled narrow valleys with moderate side slopes.

Channel materials: Bedrock bed with streambanks composed of boulders, cobble and gravel.

Slope Range: .02 - .04 (B1c < .02)   **Entrenchment Ratio:** 1.4 - 2.2

Width/depth Ratio: > 12   **Sinuosity:** > 1.2

THE MORPHOLOGICAL DESCRIPTION

B1c- Pennsylvania

B1 - North Carolina

B1c - Wyoming

THE MORPHOLOGICAL DESCRIPTION

B1- Oregon

B1 - Maryland

B1 - New Mexico

MORPHOLOGICAL DESCRIPTION AND EXAMPLES OF STREAM TYPES

B2 Stream Type

The B2 stream types are seen as moderately entrenched systems, with channel gradients of 2-4%. B2 stream types are typically located in or on coarse alluvial fans, lag deposits from old landslide debris, rockfall and talus areas, coarse colluvial deposits, and structurally controlled drainageways. Valley types that normally contain B2 stream channels include types II, III, and VII. The channel bed morphology is dominated by boulder materials, and characterized as series of rapids with irregular spaced scour pools. The B2 stream type has a moderate width/depth ratio and a sinuosity greater than 1.2. Many B2 stream channels have developed in residual materials derived from resistant rock types or from alluvial and/or colluvial deposition. The channel materials are composed primarily of boulders with lesser amounts of cobble, gravel and sand. The boulder materials are often associated with lag deposits originating from both alpine and continental glaciation. The bed and bank materials of the B2 stream types are considered stable and contribute only small quantities of sediment during runoff events.

THE MORPHOLOGICAL DESCRIPTION

DELINEATIVE CRITERIA (B2)

Landform/soils: Structural controlled narrow valleys associated with colluvium or lag deposits, narrow, moderate to gentle glaciated valleys.

Channel materials: Boulders with smaller amounts of cobble, gravel and sand.

Slope Range: .02 - .04 (B2c < .02)   **Entrenchment Ratio:** 1.4 - 2.2

Width/depth Ratio: > 12   **Sinuosity:** > 1.2

THE MORPHOLOGICAL DESCRIPTION

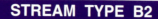
STREAM TYPE B2

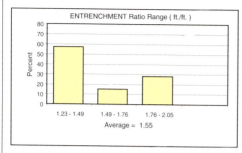

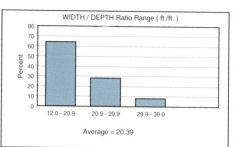

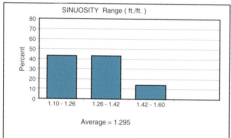

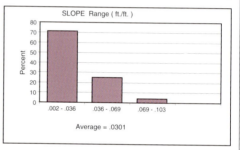

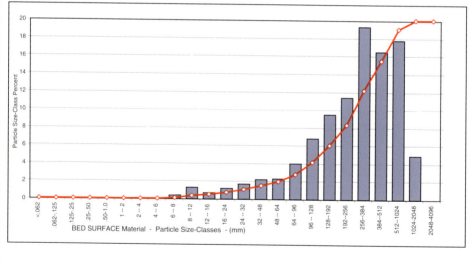

THE MORPHOLOGICAL DESCRIPTION

B2 - Idaho

B2 - California

B2 - Maryland

MORPHOLOGICAL DESCRIPTION AND EXAMPLES OF STREAM TYPES

B3 Stream Type

The B3 stream types are moderately entrenched systems with channel gradients of 2-4%. B3 stream types are typically developed in very coarse alluvial fans, lag deposits from stabilized slide debris, rockfall, talus, and very coarse colluvial deposits and structurally controlled drainage ways. Valley types that contain B3 stream channels include types II, III, and VII. The channel bed morphology is dominated by cobble materials and characterized by a series of rapids with irregular spaced scour pools. The average pool-to-pool spacing for the B3 stream type is 3-4 bankfull channel widths. Pool to pool spacing for the B3c (< 2% slope) is generally 4-5 bankfull channel widths. Pool to pool spacing adjusts inversely to stream gradient. The B3 stream type has a moderate width/depth ratio and sinuosity greater than 1.2. Many B3 stream types are associated with residual materials derived from resistant rock types or from alluvial and/or colluvial deposition. The channel materials are composed primarily of cobble with a few boulders, lesser amounts of gravel and sand. The large cobble materials often have originated from lag deposits that are the result of both alpine and continental glaciation. The bed and bank materials of the B3 stream types are stable and contribute only small quantities of sediment during runoff events. Large woody debris is an important component for fisheries habitat when available.

THE MORPHOLOGICAL DESCRIPTION

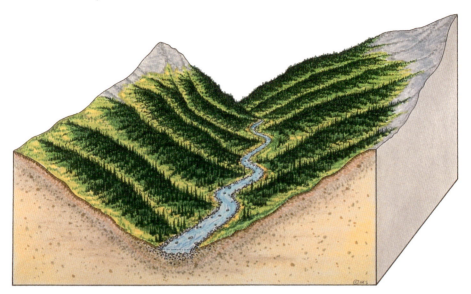

DELINEATIVE CRITERIA (B3)

Landform/soils: Narrow, moderately steep colluvial valleys with gentle side slopes. Soils are colluvium and/or alluvium. Often in fault line valleys or on well vegetated alluvial fans.

Channel materials: Predominantly cobble with lesser amounts of boulders, gravel and sand. Streambanks are stable due to coarse material.

Slope Range: .02 - .04 (B4c < .02)   **Entrenchment Ratio:** 1.4 - 2.2

Width/depth Ratio: > 12   **Sinuosity:** > 1.2

THE MORPHOLOGICAL DESCRIPTION

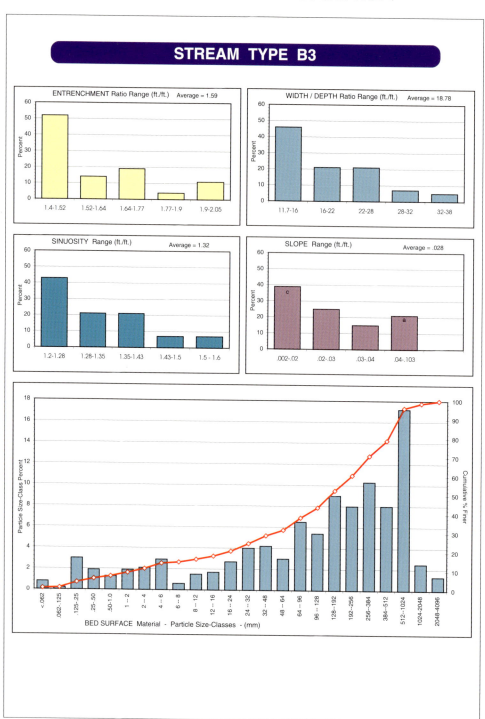

THE MORPHOLOGICAL DESCRIPTION

B3 - Wyoming

B3 - Maryland

B3 - Oregon

MORPHOLOGICAL DESCRIPTION AND EXAMPLES OF STREAM TYPES

B4 Stream Type

The B4 stream types are moderately entrenched systems on gradients of 2-4%. B4 stream types normally develop in stable alluvial fans, colluvial deposits, and structurally controlled drainage ways. Landforms are often gentle to rolling slopes in relatively narrow, colluvial or structurally controlled valleys. Valley types that contain B4 stream channels include types II, III, and VII. The channel bed morphology is dominated by gravel material and characterized as a series of rapids with irregular spaced scour pools. The average pool-to-pool spacing for the B4 stream type is 3-4 bankfull channel widths, while pool to pool spacing for the B4c (<2%) is generally 4-5 bankfull channel widths. Pool to pool spacing adjusts inversely with stream gradient. The B4 stream type has a moderate width/depth ratio and a sinuosity greater than 1.2. Many B4 stream types are associated with residual materials derived from resistant rock types or from alluvial and/or colluvial deposition. The channel materials are composed predominantly of gravel with lesser amounts of boulders, gravel, and sand. The B4 stream type is considered relatively stable and is not a high sediment supply stream channel. Large, woody, debris is an important component for fisheries habitat when available.

THE MORPHOLOGICAL DESCRIPTION

DELINEATIVE CRITERIA (B4)

Landform/soils: Narrow, moderately steep colluvial valleys, occasionally on well vegetated, stable alluvial fan, or in fault line valleys.

Channel materials: Gravel dominated with lesser amounts of boulders, cobble and sand.

Slope Range: .02 - .04 (B4c, < .02)   **Entrenchment Ratio:** 1.4 - 2.2

Width/depth Ratio: > 12   **Sinuosity:** > 1.2

THE MORPHOLOGICAL DESCRIPTION

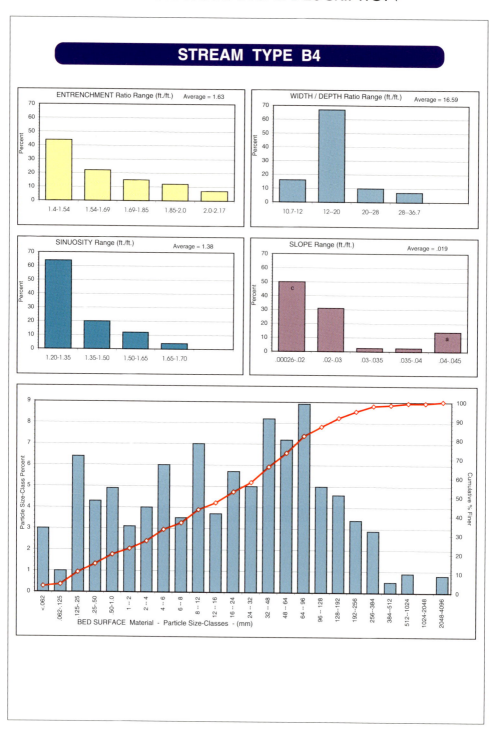

THE MORPHOLOGICAL DESCRIPTION

B4 - Ontario, Canada

B4 - Alaska

B4 - Maryland

79

MORPHOLOGICAL DESCRIPTION AND EXAMPLES OF STREAM TYPES

B5 Stream Type

The B5 stream types are moderately entrenched systems with channel gradients of 2-4%. B5 stream types are typically established on stable, well vegetated alluvial fans, colluvial deposits and relatively narrow, moderately sloping valleys. Landforms are observed as areas with gentle, rolling slopes in relatively narrow, colluvial valleys, and soils derived from residual materials including grussic granite, eolian sand deposits, and colluvial deposits. Valley types that contain B5 stream channels are types II, III, and VII. The channel bed morphology is dominated by sand-sized materials and characterized as a series of rapids with irregular spaced scour pools. The average pool-to-pool spacing for the B5 stream type is 3-4 bankfull channel widths. Pool to pool spacing for the B5c (<2% slope) is generally 4-5 bankfull channel widths. Pool to pool spacing adjusts inversely with stream gradient. The B5 stream type has a moderate width/depth ratio and a sinuosity greater than 1.2. The channel materials are composed predominantly of sand and small gravel with occasional amounts of silt/clay. The B5 stream type is relatively stable where the presence of dense riparian vegetation is noted. Large, woody, organic debris is an important component of fisheries habitat where sources are available.

THE MORPHOLOGICAL DESCRIPTION

DELINEATIVE CRITERIA (B5)

Landform/soils: Narrow, moderately steep colluvial valleys with gentle sloping side slopes. Can also be located on stable, well vegetated alluvial fans. Soils from residual, eolian, alluvium and/or colluvium.

Channel materials: Predominantly sand with lesser amounts of gravel and silt/clay.

Slope Range: .02 - .04 (B5c, < .02) **Entrenchment Ratio:** 1.4 - 2.2

Width/depth Ratio: > 12 **Sinuosity:** > 1.2

THE MORPHOLOGICAL DESCRIPTION

B5 - Colorado

B5 - Maryland

B5 - Texas

THE MORPHOLOGICAL DESCRIPTION

B5 - California

B5 - New Mexico

B5 - Idaho

MORPHOLOGICAL DESCRIPTION AND EXAMPLES OF STREAM TYPES

B6 Stream Type

The B6 stream type is a moderately entrenched system, incised in cohesive materials, with channel slopes less than 4%. The B6 stream types are found in narrow valleys containing cohesive residual soils; in depositional landscapes composed of fine, wind deposited (Loess) materials formed as gently sloping terrain; and on well vegetated alluvial fans. Valley types that contain the B6 stream channels include types II, III, and VI. The width/depth ratio of the B6 stream type is generally the lowest of all of the B stream types due to the cohesive nature of the silt/clay streambanks. B6 stream types are generally stable due to the effects of moderate entrenchment and moderate width/depth ratios. Additionally, riparian vegetation associated with the B6 type is generally very dense, except in arid environments and plays an important role in maintaining channel stability and lower width/depth ratios. These stream types are "washload" rather than "bedload" streams, and thus, have a characteristically low sediment supply and an infrequent occurrence of sediment deposition.

THE MORPHOLOGICAL DESCRIPTION

DELINEATIVE CRITERIA (B6)

Landform/soils: Narrow, moderately steep valleys with gentle sloping sideslopes. Soils either residual, alluvial and/or colluvial.

Channel materials: Silt/Clay with lesser amounts of sand.

Slope Range: .02 - .04 (B5c, < .02) **Entrenchment Ratio:** 1.4 - 2.2

Width/depth Ratio: > 12 **Sinuosity:** > 1.2

THE MORPHOLOGICAL DESCRIPTION

B6 - Montana

B6 - Texas

THE MORPHOLOGICAL DESCRIPTION

B6c - Georgia

B6c - Maryland

B6c - Wisconsin

MORPHOLOGICAL DESCRIPTION AND EXAMPLES OF STREAM TYPES

C1 Stream Type

C1 stream types are slightly entrenched, meandering, alluvial channels with bedrock controlled beds, and occur on gentle gradients in broad valleys. They are characterized by a well developed floodplain constructed of alluvium. The C1 stream type can be found in narrow, structurally controlled valleys as well as in very broad valleys and across a diversity of hydro-physiographic provinces often associated with sedimentary rock. The C1 stream channels are found in valley types IV, VI, and VIII. The C1 stream type adjusts laterally and typically has a high width/depth ratio due to the controlling influence of bedrock materials in the channel bed. The stream gradients are less than 2%. Sinuosity is moderate and meander width ratios are low relative to other alluvial streams. The spacing of pools is related to the nature and resistance of the bedrock, and backwater pools are often created by irregular spacing of large, woody organic debris. Channel materials are predominantly bedrock, although gravels and sand occur in small amounts in depositional sites. The C1 stream type is very stable. Sediment supply for this stream type is very low.

THE MORPHOLOGICAL DESCRIPTION

DELINEATIVE CRITERIA (C1)

Landform/soils: Broad, gentle gradient structural controlled alluvial valleys.
Channel materials: Bedrock bed with alluvial banks (cobble, gravel and sand).
Slope Range: < .02 (C1c- .001)   **Entrenchment Ratio:** > 2.2
Width/depth Ratio: >12   **Sinuosity:** >1.2

THE MORPHOLOGICAL DESCRIPTION

C1 - Maryland

C1 - New Mexico

C1 - North Carolina

THE MORPHOLOGICAL DESCRIPTION

C1 - West Virginia

C1 - Utah

C1 - Texas

MORPHOLOGICAL DESCRIPTION AND EXAMPLES OF STREAM TYPES

C2 Stream Type

C2 stream types are boulder-dominated, meandering, high width/depth ratio stream channels with well developed flood plains. They are generally associated with coarse lag deposits originating from extreme past flood events, or from glacial deposition. The C2 stream channel is found in valley types IV, V, VI, and VIII. Modern flows rarely entrain the dominant boulder-sized bed material. The streambanks are composed of coarse, resistant particles and in combination with the boulder materials in the bed, produce a stable stream type.

The spacing of pools is related to the nature and resistance of the boulder materials and backwater pools are often created by irregular spacing of large, woody debris. These streams have a relatively low bankfull velocity, due to the high relative roughness developing from a high width/depth ratio and large sized channel bed and bank particles. The hydraulic and roughness coefficients of the C2 channel are much different than those for the C1 stream type. Sediment supply in the C2 stream type is very low.

THE MORPHOLOGICAL DESCRIPTION

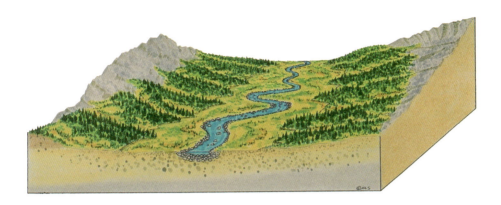

DELINEATIVE CRITERIA (C2)

Landform/soils: Broad, gentle gradient, alluvial valleys associated with lag deposits. Can also be associated with glaciated and/or structural controlled valleys.

Channel materials: Predominantly boulders with lesser amounts of cobble, gravel and sand.

Slope Range: < .02 (C2c- .001)   **Entrenchment Ratio:** > 2.2

Width/depth Ratio: >12   **Sinuosity:** >1.2

THE MORPHOLOGICAL DESCRIPTION

C2 - California

C2 - Nevada

C2 - California

THE MORPHOLOGICAL DESCRIPTION

C2 - Washington

C2 - Colorado

C2 - California

MORPHOLOGICAL DESCRIPTION AND EXAMPLES OF STREAM TYPES

C3 Stream Type

The C3 stream type is a slightly entrenched, meandering, riffle/pool, cobble-dominated channel with a well developed floodplain. The C3 stream type is found in U-shaped glacial valleys; valleys bordered by glacial and Holocene terraces; and in very broad, coarse alluvial valleys. The dominant bed material is often originating as a lag deposit from both Pleistocene and Holocene deposition and from extreme, rare floods. The C3 stream channels are found in valley types IV, V, VI, and VIII. C3 stream channels have gentle gradients of less than 2%, display a high width/depth ratio, are slightly more sinuous and have a higher meander width ratio than the C1 and C2 stream types. The riffle/pool sequence of the C3 stream type is on average at 5-7 bankfull channel widths. The streambanks are generally composed of unconsolidated, heterogenous, non-cohesive, alluvial materials that are finer than the cobble-dominated bed material. Consequently, the channel is susceptible to accelerated bank erosion. Rates of lateral adjustment are influenced by the presence and condition of riparian vegetation. Sediment supply is low, unless streambanks are in a high erodibility condition. Meander and depositional patterns which modify the condition of this stream type are described in Chapter 6.

THE MORPHOLOGICAL DESCRIPTION

DELINEATIVE CRITERIA (C3)

Landform/soils: Broad alluvial and glaciated valleys, holocene terraces generally present. Soils are glacial deposition and alluvium.

Channel materials: Predominantly cobble with lesser amounts of gravel and sand. Banks are finer in material size than channel bed.

Slope Range: < .02 (C3c- .001)

Width/depth Ratio: >12

Entrenchment Ratio: > 2.2

Sinuosity: >1.2

THE MORPHOLOGICAL DESCRIPTION

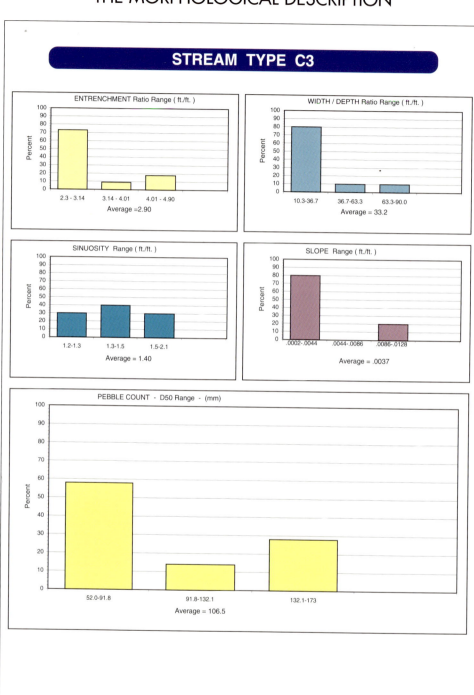

THE MORPHOLOGICAL DESCRIPTION

C3 - Montana

C3 - Arizona

C3 - California

MORPHOLOGICAL DESCRIPTION AND EXAMPLES OF STREAM TYPES

C4 Stream Type

The C4 stream type is a slightly entrenched, meandering, gravel-dominated, riffle/pool channel with a well developed floodplain. The C4 stream type is found in U-shaped glacial valleys; valleys bordered by glacial and Holocene terraces; in mountain meadows and in very broad, coarse alluvial valleys. Some of the C4 stream types occur in glacial outwash terrain, closer to the lobe where gravel material is present The C4 stream channels are found in valley types IV, V, VI, VIII, IX and X. C4 stream channels have gentle gradients of less than 2%, display a high width/depth ratio, are slightly more sinuous and have a higher meander width ratio than the C1, C2 and C3 stream types. The riffle/pool sequence for the C4 stream type average 5-7 bankfull channel widths in length. The streambanks are generally composed of unconsolidated, heterogenous, non-cohesive, alluvial materials that are finer than the gravel-dominated bed material. Consequently, the stream is susceptible to accelerated bank erosion. Rates of lateral adjustment are influenced by the presence and condition of riparian vegetation. Sediment supply is moderate to high, unless streambanks are in a very low erodibility condition. The C4 stream type, characterized by the presence of point bars and other depositional features, is very susceptible to shifts in both lateral and vertical stability caused by direct channel disturbance and changes in the flow and sediment regimes of the contributing watershed. Meander and depositional patterns which modify the condition of this stream type are described in Chapter 6.

THE MORPHOLOGICAL DESCRIPTION

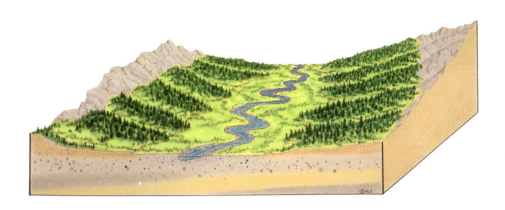

DELINEATIVE CRITERIA (C4)

Landform/soils: Broad, gentle gradient alluvial valleys and river deltas. Soils are alluvium.
Channel materials: Predominantly gravel, with lesser amounts of cobble, sand and silt/clay.
Slope Range: < .02 (C4c- .001)   **Entrenchment Ratio:** > 2.2
Width/depth Ratio: >12   **Sinuosity:** >1.2

THE MORPHOLOGICAL DESCRIPTION

STREAM TYPE C4

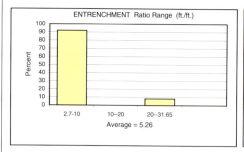
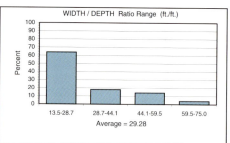
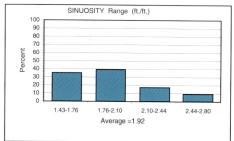
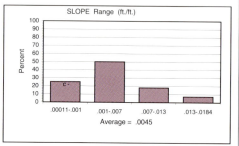
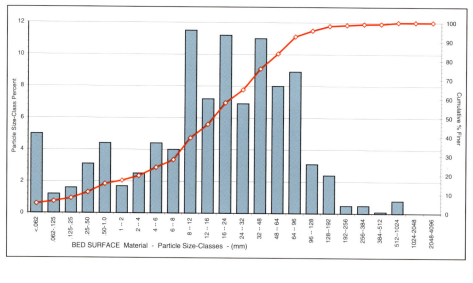

THE MORPHOLOGICAL DESCRIPTION

C4 - New Mexico

C4 - Colorado

C4 - Wisconsin

MORPHOLOGICAL DESCRIPTION AND EXAMPLES OF STREAM TYPES

C5 Stream Type

California

The C5 stream type is a slightly entrenched, meandering, sand-dominated, riffle/pool channel with a well developed floodplain. The C5 stream type occurs in broad valleys and plains areas with a history of riverine, lacustrine, glacial (outwash and glacio-fluvial), and eolian deposition. The C5 stream type can be found in very low relief basins typical of the interior lowlands, great plains, coastal plains, and in river deltas. Glacial outwash areas can also develop C5 stream types. The C5 stream channels are found in valley types IV, V, VI, VIII, IX, X, and XI. It is obvious that the C5 stream type is widely distributed throughout a wide range of physiographic provinces. Generally, C5 stream channels have gentle gradients of less than 2%. Gradients less than 0.001 are denoted as a C5c- to indicate the slope condition of many C5 stream types. The C5 stream channel displays a higher width/depth ratio than the C4 and C3 stream types due to the depositional characteristic of the stream bed and the active lateral migration tendencies. The riffle/pool sequence for the C5 stream type averages 5-7 bankfull channel widths in length. Bed forms of ripples, dunes, and anti-dunes are prevalent. The streambanks are generally composed of sandy material, with stream beds exhibiting little difference in pavement and sub-pavement material composition. Rates of lateral adjustment are influenced by the presence and condition of riparian vegetation. Sediment supply is high to very high, unless streambanks are in a very low erodibility condition. The C5 stream type, characterized by the presence of point bars and other depositional features, is very susceptible to shifts in both lateral and vertical stability caused by direct channel disturbance and changes in the flow and sediment regimes of the contributing watershed. Meander and depositional patterns which modify the condition of this stream type are described in Chapter 6.

THE MORPHOLOGICAL DESCRIPTION

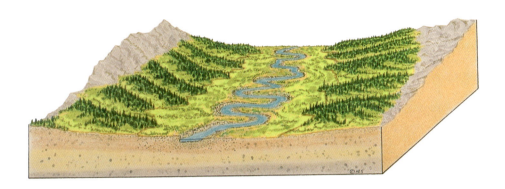

DELINEATIVE CRITERIA (C5)

Landform/soils: Broad, gentle gradient alluvial valleys, river deltas, broad plains. Soils are depositional such as lacustrine, glacial outwash, eolian.

Channel materials: Predominantly sand bed and banks, with occasional gravel and silt/clay. Streambanks may contain finer particles than bed material.

Slope Range: < .02 (C5c- .001)
Width/depth Ratio: >12
Entrenchment Ratio: > 2.2
Sinuosity: >1.2

THE MORPHOLOGICAL DESCRIPTION

STREAM TYPE C5

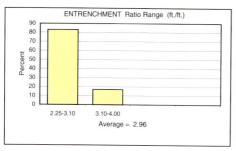
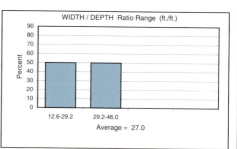
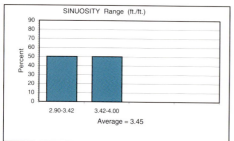
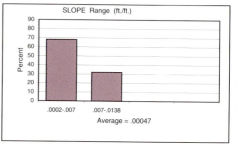
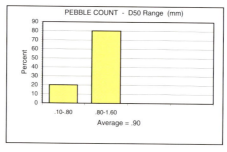

106

THE MORPHOLOGICAL DESCRIPTION

C5 - Colorado

C5 - Colorado

C5 - Montana

MORPHOLOGICAL DESCRIPTION AND EXAMPLES OF STREAM TYPES

C6 Stream Type

The C6 stream type is a slightly entrenched, meandering, silt-clay dominated, riffle-pool channel with a well developed floodplain. The C6 stream type occurs in broad valleys and plains areas with a history of riverine, lacustrine, and eolian deposition (loess). The C6 stream type can be found in very low relief basins typical of the interior lowlands, great plains, coastal plains, and in river deltas. The lower extremities of glacial outwash areas can also develop C6 stream types. The C6 stream channels are associated with valley types IV, V, VI, VIII, IX, X, and XI. It is obvious that the C6 stream type is widely distributed throughout a wide range of physiographic provinces. Generally, C6 stream channels have gentle gradients of less than 2%. Gradients less than 0.001 are denoted as a C6c- to indicate the very low gradients of many C6 stream types. The C6 stream channel displays a lower width/depth ratio than all of the other C stream types due to the cohesive nature of stream bank materials. The riffle/pool sequence for the C6 stream type is, on average, 5-7 bankfull channel widths in length. The streambanks are generally composed of silt-clay and organic materials, with the stream beds exhibiting little difference in pavement and sub-pavement material composition. Rates of lateral adjustment are influenced by the presence and condition of riparian vegetation. Sediment supply is moderate to high, unless streambanks are in a very high erodibility condition. Bedload sediment yields for the stream types are typically low, reflecting the presence of fine bed and bank materials and gentle channel slopes. The C6 stream type is very susceptible to shifts in both lateral and vertical stability caused by direct channel disturbance and changes in the flow and sediment regimes of the contributing watershed. Meander and depositional patterns which modify the condition of this stream type are described in Chapter 6.

THE MORPHOLOGICAL DESCRIPTION

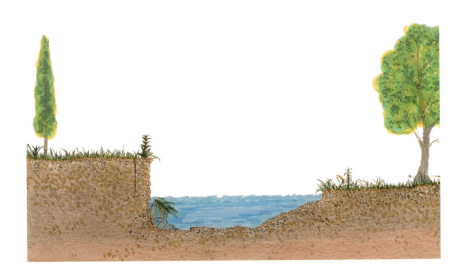

DELINEATIVE CRITERIA (C6)

Landform/soils: Broad gentle valleys, plains, and deltas. Depositional soils (alluvium), associated with cohesive materials from riverine and lacustrine process. Often associated with tidal influence deltas, marshes and other wetland complexes.

Channel materials: Silt-clay predominates, however many of these C6 stream types are associated with a high organic component including peat.

Slope Range: < .02 (C6c- .001) (.0001 more common) **Entrenchment Ratio:** > 2.2

Width/depth Ratio: >12 (generally lowest of C's) **Sinuosity:** >1.2

THE MORPHOLOGICAL DESCRIPTION

C6 - Colorado

C6 - Texas

C6 - Georgia

THE MORPHOLOGICAL DESCRIPTION

C6 - Georgia

C6 - Texas

C6 - Texas

MORPHOLOGICAL DESCRIPTION AND EXAMPLES OF STREAM TYPES

D3 Stream Type

The D3 stream types are multiple channel systems, described as braided streams found within broad alluvial valleys and on alluvial fans consisting of coarse depositional materials formed into moderately steep terrain. Primarily, the braided system consists of interconnected distributary channels formed in depositional environments. The D3 stream type occurs in moderately steep, narrow, U-shaped glacial valleys; on alluvial fans; and in gentle gradient alluvial valleys. The D3 stream channels may be found in valley types III, V, and VIII. Channel bed materials are predominantly cobble, with a strong bi-modal distribution of sands. Gravel is also present, but often imbedded with sands. The braided channel system is characterized by high bank erosion rates, excessive deposition occurring as both longitudinal and transverse bars, and annual shifts of the bed locations. Bed features are developed from convergence/divergence processes. The channels generally are at the same gradient as their parent valley. A combination of adverse conditions are responsible for channel braiding, including high sediment supply, high bank erodibility, moderately steep gradients, and very flashy runoff conditions which can vary rapidly from a base flow to an over-bank flow on a frequent basis. Characteristic width/depth ratios are very high, exceeding values of 40 to 50, with values of 400 or larger often noted. D3 channel gradients are generally less than 2%; however, D3 types can also develop within alluvial fans which have slopes of 2% to 4% (D3b). Observations have been made of braided streams on alluvial fans with slopes greater than 4% (D3a). The D3 is a very high sediment supply system and typically produces high sediment yields.

THE MORPHOLOGICAL DESCRIPTION

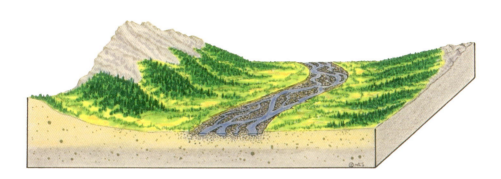

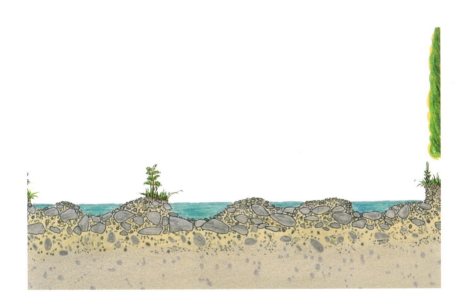

DELINEATIVE CRITERIA (D3)

Landform/soils: Moderately steep glacial valleys, alluvial fans, narrow fluvial mountain valleys and terraced valleys in coarse alluvium. Bed material can be lag deposit.

Channel materials: Cobble dominated with a mixture of gravel and sand. Bank materials are finer than bed material, and generally actively eroding.

Slope Range: < .04 **Entrenchment Ratio:** NA (not incised)

Width/depth Ratio: > 40 **Sinuosity:** Low, channel slope = valley slope

THE MORPHOLOGICAL DESCRIPTION

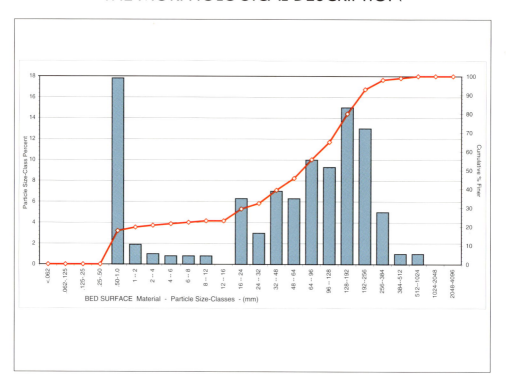

D3 - Montana

THE MORPHOLOGICAL DESCRIPTION

D3 - Colorado

D3 - Montana

D3 - Colorado

115

MORPHOLOGICAL DESCRIPTION AND EXAMPLES OF STREAM TYPES

D4 Stream Type

Colorado

The D4 stream types are multiple channel systems, described as braided streams, found within broad alluvial valleys and on alluvial fans consisting of coarse depositional materials formed into moderately steep terrain. Primarily, the braided system consist of interconnected distributary channels formed in depositional environments. The D4 stream type occurs in moderately steep, narrow, U-shaped glacial valleys; on alluvial fans; and in gentle gradient alluvial valleys. This stream type can also occur on low relief river deltas, as well as on the upper lobe of glacial outwash valleys. The D4 stream channels may be found in valley types III, V, VIII, IX, X, and XI. Channel bed materials are predominantly gravel, with a strong bi-modal distribution of sands. Cobble may be found in lesser amounts, often imbedded with sands. The braided channel system is characterized by high bank erosion rates, excessive deposition occurring as both longitudinal and transverse bars, and annual shifts of the bed locations. Bed morphology is characterized by a closely spaced series of rapids and scour pools formed by convergence/divergence processes that are very unstable. The channels generally are at the same gradient as their parent valley. A combination of adverse conditions are responsible for channel braiding, including high sediment supply, high bank erodibility, moderately steep gradients, and very flashy runoff conditions which can vary rapidly from a base flow to an over-bank high flow on a frequent basis. Characteristic width/depth ratios are very high, exceeding values of 40 to 50 with values of 400 or larger often noted. D4 channel gradients are generally less than 2%; however, D4 types can also develop within alluvial fans which have slopes of 2% to 4% (D4b). Observations have been made of braided streams on alluvial fans with slopes greater than 4% (D4a). The D4 is a very high sediment supply system, and typically produces high bedload sediment yields.

THE MORPHOLOGICAL DESCRIPTION

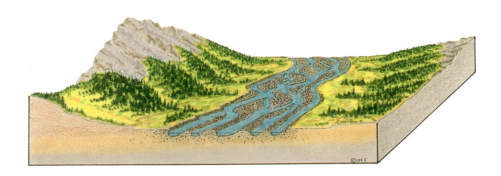

DELINEATIVE CRITERIA (D4)

Landform/soils: Moderately steep glacial valleys, alluvial fans, narrow fluvial mountain valleys and terraced valleys in coarse alluvium. Can occur in gravel splays, and coarse delta deposits.

Channel materials: Gravel bed with smaller quantities of cobble. Typical is a bi-modal distribution of sands. Stream bank materials generally finer than bed, actively eroding.

Slope Range: < .04   **Entrenchment Ratio:** N/A (not incised)

Width/depth Ratio: > 40   **Sinuosity:** Low, channel slope = valley slope

THE MORPHOLOGICAL DESCRIPTION

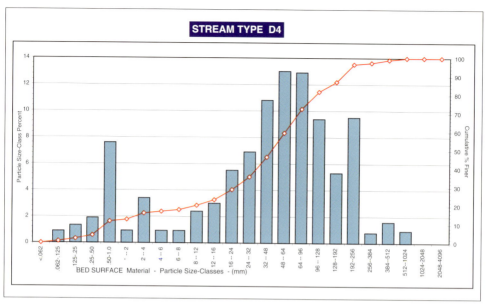

D4 - Colorado

D4 - Colorado

THE MORPHOLOGICAL DESCRIPTION

D4 - Montana

D4 - Montana

D4 - Colorado

MORPHOLOGICAL DESCRIPTION AND EXAMPLES OF STREAM TYPES

D5 Stream Type

The D5 stream types are multiple channel systems described as braided streams, found within broad alluvial valleys and on alluvial fans consisting of deposited sand-sized materials. The braided system consists of interconnected distributary channels formed in depositional environments. The D5 stream type occurs in gentle gradient, narrow, U-shaped glacial valleys consisting of glacio-lacustrine deposits, sand dunes (eolian); in very low relief alluvial valleys; and in glacial outwash areas and deltas. The D5 stream channels may be found in valley types III, V, VIII, IX, X, and XI. Channel bed materials are predominantly sand, with interspersed amounts of silt/clay materials on deltas and in varves of lacustrine depositional areas. The braided channel system is characterized by high bank erosion rates, excessive deposition occurring as both longitudinal and transverse bars, and annual shifts of the bed location. Bed morphology is characterized by a closely spaced series of rapids and scour pools formed by convergence/divergence processes that are very unstable. The channels generally are of the same gradient as their parent valley. A combination of adverse conditions are responsible for channel braiding, including high sediment supply, high bank erodibility, moderately steep gradients, and very flashy runoff conditions which can vary rapidly from a base flow to an over-bank flow on a frequent basis. Characteristic width/depth ratios are very high, exceeding values of 40 to 50 with values of 400 or larger often noted. D5 channel gradients are generally less than 2%; however, D5 types can also develop within alluvial fans which have slopes of 2% to 4% (D5b). Observations have been made of braided streams on alluvial fans with slopes greater than 4% (D5a). The D5 is a very high sediment supply system, and typically produces high bedload sediment yields.

THE MORPHOLOGICAL DESCRIPTION

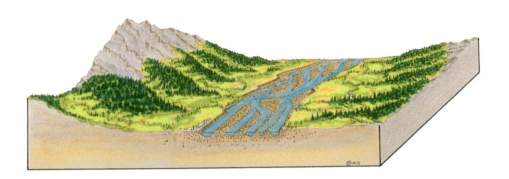

DELINEATIVE CRITERIA (D5)

Landform/soils: Moderately steep to very gentle alluvial valleys, glacial outwash plains, eolian dunes and deltas. Soils are depositional from riverine, lacustrine and eolian processes.

Channel materials: Sand dominated with occasional gravel and silt/clay.

Slope Range: < .04

Width/depth Ratio: > 40

Entrenchment Ratio: N/A (not incised)

Sinuosity: Low, channel slope = valley slope

THE MORPHOLOGICAL DESCRIPTION

D5 - Wyoming

D5 - Colorado

D5 - Colorado

THE MORPHOLOGICAL DESCRIPTION

D5 - Colorado

D5 - Texas

D5 - Montana

MORPHOLOGICAL DESCRIPTION AND EXAMPLES OF STREAM TYPES

D6 Stream Type

Alaska

The D6 stream types are multiple channel systems described as braided streams, found within broad alluvial valleys and in river deltas consisting of cohesive silt-clay depositional materials. The braided system consists of interconnected distributary channels formed in depositional environments. The D6 stream type occurs typically in wide valleys with very gentle gradients. Lacustrine deposits and river deltas are the most typical landform/materials. The D5 stream channels are found in valley types VIII, X and XI. Channel bed materials are predominantly silt-clay with a contribution of organic material such as peat. The braided channel system is characterized by high bank erosion rates, excessive deposition occurring as both longitudinal and transverse bars, and annual shifts of the bed location. Bed morphology is characterized by a closely spaced series of rapids and scour pools formed by convergence/divergence processes that are very unstable. The channels generally are of the same gradient as their parent valley. A combination of adverse conditions are associated with braiding, including high sediment supply, high bank erodibility, moderately steep gradients, and very flashy runoff conditions which can vary rapidly from a base flow to an over-bank flow on a frequent basis. Characteristic width/depth ratios are very high, exceeding values 40 to 50, with values of 400 or larger often noted. D6 channel gradients are generally less than .001% slope. The D6 is a high sediment supply system that typically produces low bedload sediment yields and high suspended or washload sediment yields.

THE MORPHOLOGICAL DESCRIPTION

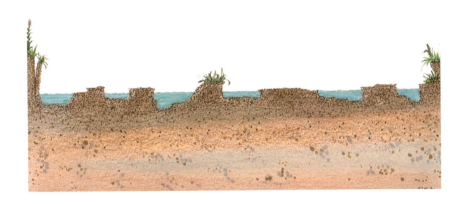

DELINEATIVE CRITERIA (D6)

Landform/soils: Gentle, wide alluvial valleys, Lacustrine deposits, and deltas. Cohesive depositional soils. Organic lenses and peat common in deltas and some lacustrine features.

Channel materials: Silts and/or clays dominate the D6 stream type. Organic contributions are common such as peat.

Slope Range: < .02 (generally less than .0001) **Entrenchment Ratio:** N/A (not incised)

Width/depth Ratio: > 40 **Sinuosity:** Low, channel slope = valley slope

MORPHOLOGICAL DESCRIPTION AND EXAMPLES OF STREAM TYPES

DA Stream Type

British Columbia

The DA stream types are highly interconnected channel systems developing in gentle relief terrain areas consisting of cohesive soil materials and exhibiting wetland environments with stable channel conditions. Landforms are seen as unconfined, broad valleys with well developed floodplains, or delta areas that are more typical of marshes with stable channels. The DA stream type is characteristic of a balance between the rate of basin filling and basin subsidence such that an equilibrium condition is maintained over time. Valley types supporting the DA stream systems are types X and XI. The delta type XI is considered as a stable delta system, supporting wetlands such as fresh water marshes and tidal influenced salt marshes, with system base levels controlled by lakes or sea level, to maintain stable elevations. In broad alluvial valleys, the anastomosed channels are generally stable due to presence of cohesive bank materials and extensively developed riparian vegetation. The stream beds are often vertically accreting, but kept in balance due to the subsidence effects of tectonically active basins (Smith and Putnam, 1980). The anastomosed channel patterns typically display a range of low to high width/depth ratios and a similar range of sinuosities, with high meander belt widths. Channel slopes are less than 0.5% but typically in the .01% range. Dominant channel bed particle sizes for this stream type vary from gravel (DA4), to sand (DA5), to silt-clay (DA6). Most of the channel banks, however, contain a highly cohesive material component, intermixed with a dense root mass. Peat is commonly found. Sediment supply and bedload contributions are low, as sediment transport is dominated by wash load or suspended sediment.

THE MORPHOLOGICAL DESCRIPTION

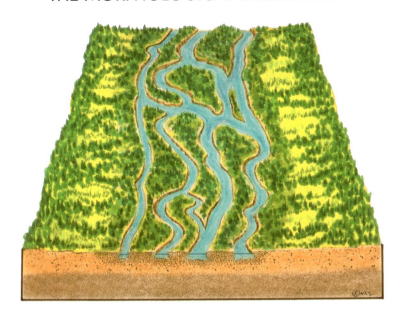

DELINEATIVE CRITERIA (DA3-DA6))

Landform/soils: Broad gentle valleys and deltas. Wetland environments with stable islands, often cohesive banks mixed with organic material. Depositional soils.

Channel materials: The materials vary from gravel (DA4), sand (DA5), to silt/clay (DA6). Peat and other organic materials are very common with these streams.

Slope Range: < .005 (average closer to .0001)   **Entrenchment Ratio:** N/A (not incised)

Width/depth Ratio: Highly variable   **Sinuosity:** Highly variable

THE MORPHOLOGICAL DESCRIPTION

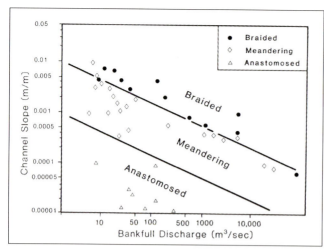

Relations of channel patterns versus slope and bankfull discharge. (Smith and Putnam, 1980).

DA - Anastomosed river.

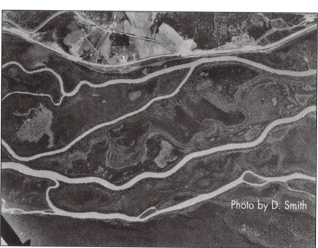

DA - British Columbia. Anastomosed river channels.

THE MORPHOLOGICAL DESCRIPTION

DA - Anastomosed river.

DA - Anastomosed river.

DA - Anastomosed river.

MORPHOLOGICAL DESCRIPTION AND EXAMPLES OF STREAM TYPES

E3 Stream Type

The E3 stream types are seen as systems with moderate sinuosities, with gentle to moderately steep channel gradients, and with very low width/depth ratios. The E stream types are located in a variety of land forms including high mountain meadows, alpine tundra, and broad alluvial valleys with well developed floodplains. The E3 stream channels are found in valley types VIII and X. The E3 stream type exhibits predominantly cobble-sized bed materials, with channel slopes less than 2 %; however, they can also develop with slopes of 2-4% (E3b) and in some cases with slopes greater than 4% (E3a). The E3 stream type can develop with a wide range of channel slopes due to the nature of the inherently stable bed and banks. Sinuosities and meander width ratios decrease, however, with an increase in channel slope.

Streambanks are composed of materials finer than that of the dominant channel bed materials, and are typically stabilized with dense riparian or wetland vegetation that forms densely rooted sod mats from grasses and grass-like plants as well as woody species. Typically the E3 stream channels have high meander width ratios, high sinuosities, and low width/depth ratios. The E3 stream types are hydraulically efficient channel forms, and they maintain a high sediment transport capacity. The narrow and relatively deep channels maintain a high resistance to plan form adjustment, which results in channel stability without significant downcutting. The E3 stream channels are very stable unless the streams are disturbed, and significant changes in sediment supply and/or streamflow occur.

THE MORPHOLOGICAL DESCRIPTION

DELINEATIVE CRITERIA (E3)

Landform/soils: Broad, gentle to moderately steep alluvial valleys.

Channel materials: Cobble dominated with fewer accumulations of gravel and sand. Stream banks have gravel/sand matrix mixed with dense root mats/organic material. Very stable.

Slope Range: < .02 (E3b, .02-.04)   **Entrenchment Ratio:** > 2.2

Width/depth Ratio: < 12   **Sinuosity:** > 1.5 (less if E3b)

THE MORPHOLOGICAL DESCRIPTION

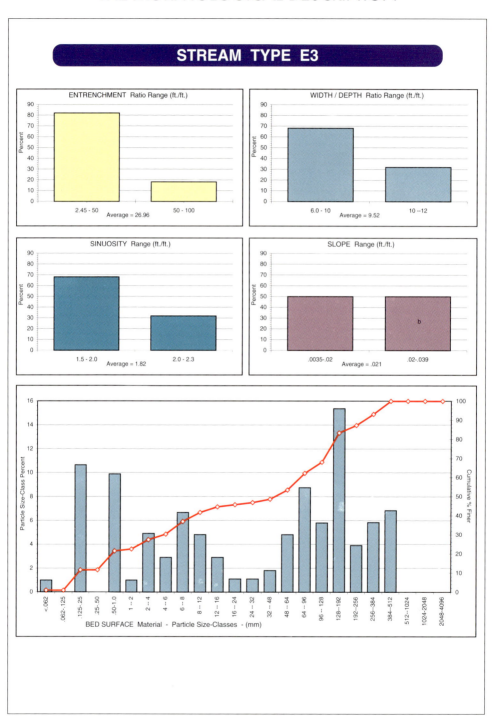

THE MORPHOLOGICAL DESCRIPTION

E3 - Wyoming

E3 - Colorado

E3 - Colorado

MORPHOLOGICAL DESCRIPTION AND EXAMPLES OF STREAM TYPES

E4 Stream Type

The E4 stream types are channel systems with low to moderate sinuosities, gentle to moderately steep channel gradients, with very low channel width/depth ratios. The E4 type is a riffle/pool stream found in a variety of land forms including high mountain meadows, alpine tundra, deltas, and broad alluvial valleys with well developed floodplains. The E4 stream channels are found in valley types VIII, X, and XI. The E4 stream type exhibits predominantly gravel-sized bed materials, with channel slopes less than 2%; however, they can also develop with slopes of 2-4% (E4b). Due to the inherently stable nature of the bed and banks, this stream type can develop with a wide range of channel slopes. Sinuosities and meander width ratios decrease, however, with an increase in slope. Streambanks are composed of materials finer than that of the dominant channel bed materials, and are typically stabilized with extensive riparian or wetland vegetation that forms densely rooted sod mats from grasses and grass-like plants, as well as woody species. Typically the E4 stream channels have high meander width ratios, high sinuosities, and low width/depth ratios. The E4 stream types are hydraulically efficient channel forms and they maintain a high sediment transport capacity. The narrow and relatively deep channels maintain a high resistance to plan form adjustment which results in channel stability without significant downcutting. The E4 stream channels are very stable unless the stream banks are disturbed, and significant changes in sediment supply and/or streamflow occur.

THE MORPHOLOGICAL DESCRIPTION

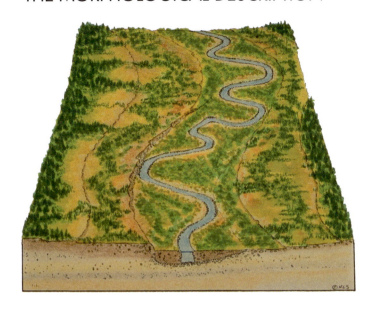

DELINEATIVE CRITERIA (E4)

Landform/soils: Gentle slopes in broad riverine or lacustrine valleys and river deltas.

Channel materials: Gravel dominated bed with smaller accumulations of sand and occasional cobble. Streambanks composed of sandy/gravel mixture with dense root mat.

Slope Range: < 0.02   **Entrenchment Ratio:** > 2.2

Width/depth Ratio: < 12   **Sinuosity:** > 1.5

THE MORPHOLOGICAL DESCRIPTION

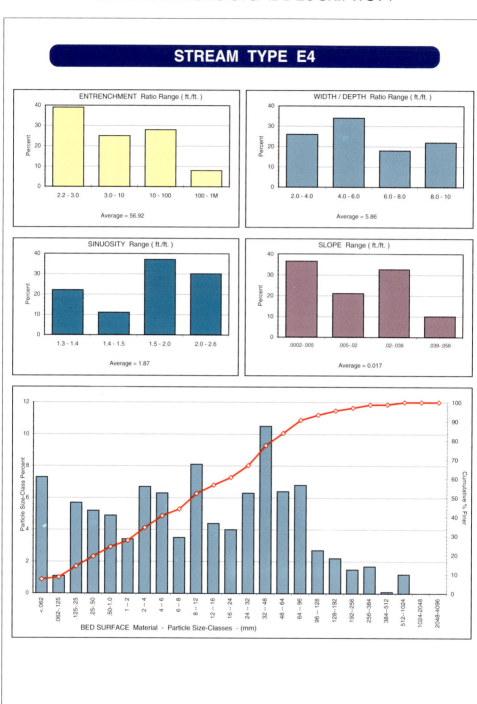

THE MORPHOLOGICAL DESCRIPTION

E4 - Wyoming

E4 - Colorado

E4 - Wyoming

137

MORPHOLOGICAL DESCRIPTION AND EXAMPLES OF STREAM TYPES

E5 Stream Type

Colorado

The E5 stream types are channel systems with low to moderate sinuosities, gentle to moderately steep channel gradients, and very low channel width/depth ratios. The E5 stream type is found in a variety of land forms including high mountain meadows, alpine tundra, deltas, lacustrine valleys, and broad alluvial valleys with well developed floodplains. The E5 stream channels are found in valley types VIII, X, and XI. The E5 stream type is typically seen as a riffle/pool system with channel slopes less than 2%. Due to the inherently stable nature of the bed and banks, this stream type can develop with a wide range of channel slopes. Sinuosities and meander width ratios decrease, however, with an increase in slope. Streambanks are composed of materials finer than that of the dominant channel bed materials, and are typically stabilized with extensive riparian or wetland vegetation that forms densely rooted sod mats from grasses and grass-like plants, as well as woody species. Typically the E5 stream channel has high meander width ratios, high sinuosities, and low width/depth ratios. The E5 stream types are hydraulically efficient channel forms and they maintain a high sediment transport capacity. The narrow and relatively deep channels maintain a high resistance to plan form adjustment which results in channel stability without significant downcutting. The E5 stream channels are very stable unless the streambanks are disturbed, and significant changes in sediment supply and/or streamflow occur.

THE MORPHOLOGICAL DESCRIPTION

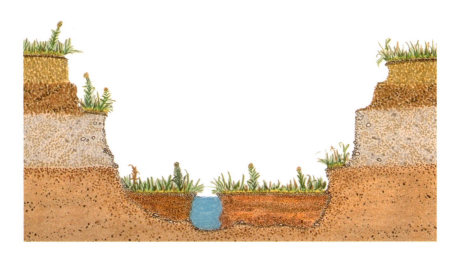

DELINEATIVE CRITERIA (E5)

Landform/soils: Gentle slopes in broad riverine or lacustrine valleys and river deltas. Can be laterally contained in entrenched valley, evolving to a channel inside a previous channel.

Channel materials: Sand dominated bed with smaller accumulations of gravel and occasional silt/clay. Streambanks composed of sandy/silt/clay mixture with dense root mat.

Slope Range: < 0.02 **Entrenchment Ratio:** > 2.2

Width/depth Ratio: < 12 **Sinuosity:** > 1.5

THE MORPHOLOGICAL DESCRIPTION

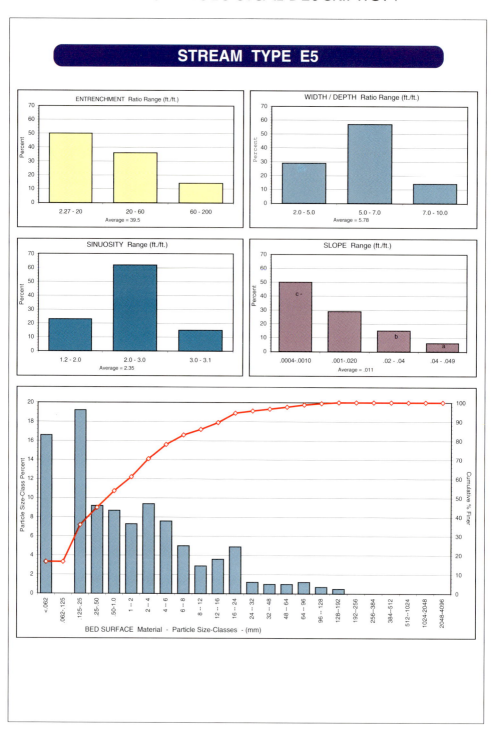

THE MORPHOLOGICAL DESCRIPTION

E5 - Colorado

E5 - Wisconsin

E5 - Colorado

141

MORPHOLOGICAL DESCRIPTION AND EXAMPLES OF STREAM TYPES

E6 Stream Type

Maryland

The E6 stream types are channel systems with low to moderate sinuosities, gentle to moderately steep channel gradients, and very low channel width/depth ratios. The E6 stream types are found in a variety of land forms including high mountain meadows, alpine tundra, deltas, lacustrine valleys, and broad alluvial valleys with well developed floodplains. The E6 stream channels are found in valley types VIII, X, and XI. The E6 stream type is typically seen as a riffle/pool system with the dominant channel materials composed of silt-clay, interspersed with organic materials. Channel slopes are less than 2%, with a high number having slopes of less than .01%. Due to the inherently stable nature of the bed and banks, this stream type can exist on a wide range of slopes. Sinuosities and meander width ratios decrease, however, with an increase in slope.

Streambanks are composed of materials similar to those of the dominant bed materials and are typically stabilized with riparian or wetland vegetation that forms densely rooted sod mats from grasses and grass-like plants as well as woody species. Typically the E6 stream channel has high meander width ratios, high sinuosities, and low width/depth ratios. The E6 stream types are hydraulically efficient forms as they require the least cross-sectional area per unit of discharge. The narrow and relatively deep channels maintain a high resistance to plan form adjustment which results in channel stability without significant downcutting. The E6 stream channels are very stable unless the streambanks are disturbed and significant changes in sediment supply and/or streamflow occur.

THE MORPHOLOGICAL DESCRIPTION

DELINEATIVE CRITERIA (E6)

Landform/soils: Gentle slopes in broad riverine or lacustrine valleys and river deltas. Can be laterally contained in entrenched valley, evolving to a channel inside a previous channel.

Channel materials: Silt/clay dominated cohesive channel materials with accumulations of organic material including peat. Dense root mat on streambanks.

Slope Range: < .02 (often < .0001)   **Entrenchment Ratio:** > 2.2

Width/depth Ratio: < 12   **Sinuosity:** > 1.5

THE MORPHOLOGICAL DESCRIPTION

E6 - Colorado

E6 - Nevada

E6 - Colorado

THE MORPHOLOGICAL DESCRIPTION

E6 - Texas

E6 - Utah

E6 - Nevada

MORPHOLOGICAL DESCRIPTION AND EXAMPLES OF STREAM TYPES

F1 Stream Type

Ontario, Canada

The F1 stream type is an entrenched, meandering, high width/depth ratio channel that is deeply incised in valleys that are structurally controlled with bedrock. The F1 stream channels are often entrenched in highly weathered rock formations configured as low relief landforms with low valley gradients. Side slopes of the F1 stream types are often vertical and confine the river laterally for great distances. The F1 stream channels are located in valley types IV and VI. The F1 stream type does not have developed floodplains, with all of the natural range of flows contained in a similar width channel. The dominant channel materials are principally bedrock, with boulders, cobble, and gravel present in fewer quantities. The F1 stream type has a relatively low to moderate sinuosity and low meander width ratios due to the degree of natural entrenchment and lateral containment. The "top of banks" of this stream type cannot be reached by floods that may be developed with the modern-day climate. The F1 stream type typically exhibits low sediment deposition, due to the low sediment supply from the relatively stable bed and banks. These systems are considered very stable stream types due to the resistant nature of their channel materials, and basically have not changed or significantly adjusted in modern times.

THE MORPHOLOGICAL DESCRIPTION

DELINEATIVE CRITERIA (F1)

Landform/soils: The F1 stream type is associated with deeply entrenched, structurally controlled, gentle gradient valleys and gorges. The F1 stream type is associated with highly weathered bedrock in a combination of river downcutting and uplift of valley walls.

Channel materials: Bedrock dominated channel with accumulations of boulders, cobble and gravel. Some sand deposits in pools and backwater eddies.

Slope Range: < .02   **Entrenchment Ratio:** < 1.4

Width/depth Ratio: > 12   **Sinuosity:** > 1.2

THE MORPHOLOGICAL DESCRIPTION

F1 - Texas

F1 - Colorado

F1 - Texas

THE MORPHOLOGICAL DESCRIPTION

F1 - New Mexico

F1 - Texas

MORPHOLOGICAL DESCRIPTION AND EXAMPLES OF STREAM TYPES

F2 Stream Type

The F2 stream type is an entrenched, meandering, high width/depth ratio channel that is deeply incised in valleys that are structurally controlled with boulder materials. The F2 stream channels are often entrenched in highly weathered rock formations configured as low relief landforms with low valley gradients. Side slopes of the F2 stream types are often vertical and confine the river laterally for great distances. The F2 stream channels are located in valley types IV and VI. The F2 stream type does not have developed floodplains, with all of the natural range of flows contained in a similar width channel. The dominant channel materials are boulders, with cobble and gravel present in fewer quantities. The F2 stream type has a relatively low to moderate sinuosity and low meander width ratios due to the degree of natural entrenchment and lateral containment. The "top of banks" of this stream type cannot be reached by floods that may be developed with the modern-day climate. The F2 stream type typically exhibits low sediment deposition due to the low sediment supply from the relatively stable bed and banks. These systems are considered very stable stream types due to the resistant nature of their channel materials, and basically have not changed or significantly adjusted in modern times.

THE MORPHOLOGICAL DESCRIPTION

DELINEATIVE CRITERIA (F2)

Landform/soils: The F2 stream type is associated with deeply entrenched, structurally controlled, gentle gradient valleys and gorges. The F2 stream type is associated with highly weathered bedrock in a combination of river downcutting and uplift of valley walls.

Channel materials: Boulder dominated channel with accumulations of cobble and gravel. Some sand deposits in pools and backwater eddies.

Slope Range: < .02   **Entrenchment Ratio:** < 1.4

Width/depth Ratio: > 12   **Sinuosity:** > 1.2

THE MORPHOLOGICAL DESCRIPTION

F2 - Utah

F2 - Texas

F2 - California

THE MORPHOLOGICAL DESCRIPTION

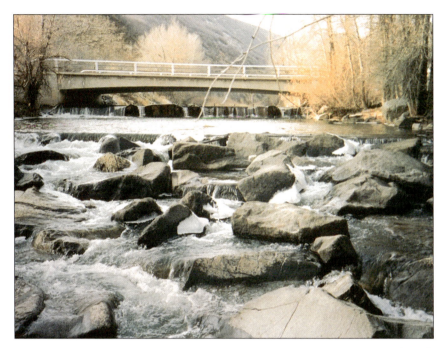

F2 - Utah

F2 - Montana

MORPHOLOGICAL DESCRIPTION AND EXAMPLES OF STREAM TYPES

F3 Stream Type

Idaho

The F3 stream type is a cobble dominated, entrenched, meandering channel, deeply incised in gentle terrain. The "top of banks" elevation for this stream type is much greater than the bankfull stage, which is indicative of the deep entrenchment. The F3 stream type can be incised in alluvial valleys, resulting in the abandonment of former floodplains. The F3 stream channels are found in valley types IV, VI, VIII, and X. The F3 channels have slopes that are generally less than 2%, exhibit riffle/pool bed features, and have width/depth ratios that are high to very high. The dominant channel materials are cobble, with lesser accumulations of gravel and sand. Often the sands will be imbedded with the large cobble sizes. Sediment supply in the F3 stream types is moderate to high, depending on bank erodibility conditions. Depositional features (central and transverse bars) are common, and related to the high sediment supply from streambanks and the high width/depth ratios. Riparian vegetation plays a marginal role in streambank stability due to the typically very high bank heights which extend beyond the rooting depth of riparian plants. Exceptions to this are the F3 stream types in the Northeast, Northwest, and Southeast United States where the relatively longer growing seasons and ample precipitation results in the establishment of riparian vegetation that tends to cover the entire slope facet of channel banks.

THE MORPHOLOGICAL DESCRIPTION

DELINEATIVE CRITERIA (F3)

Landform/soils: The F3 stream type is associated with deeply entrenched, structurally controlled, gentle gradient valleys and gorges. The F3 stream type is associated with highly weathered bedrock or depositional soils involving a combination of river downcutting and uplift of valley walls.

Channel materials: Cobble dominated channel with accumulations of gravel and sand. Streambanks generally gravel/sand matrix and unstable unless well vegetated.

Slope Range: < .02   **Entrenchment Ratio:** < 1.4

Width/depth Ratio: > 12   **Sinuosity:** > 1.2

STREAM TYPE F3

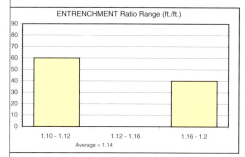
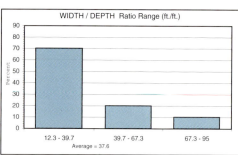
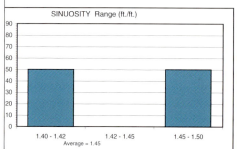
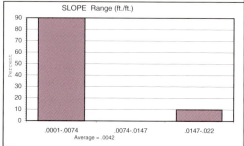
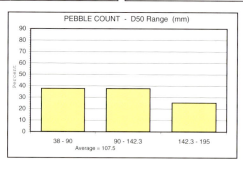

THE MORPHOLOGICAL DESCRIPTION

F3 - North Carolina

F3 - Arizona

F3 - Colorado

MORPHOLOGICAL DESCRIPTION AND EXAMPLES OF STREAM TYPES

F4 Stream Type

New Mexico

The F4 stream type is a gravel dominated, entrenched, meandering channel, deeply incised in gentle terrain. The "top of banks" elevation for this stream type is much greater than the bankfull stage, which is indicative of the deep entrenchment. The F4 stream type can be incised in alluvial valleys, resulting in the abandonment of former floodplains. The F4 stream channels are found in valley types IV, VI, VIII, X, and XI. The F4 channels have slopes that are generally less than 2%, exhibit riffle/pool bed features, and have width/depth ratios that are high to very high. The dominant channel materials are gravel, with lesser accumulations of cobble and sands. Often the sand will be imbedded with the cobble and gravel. Sediment supply in the F4 stream types is moderate to high, depending on stream bank erodibility conditions. Depositional features are common in this stream type, and over time, tend to promote development a flood plain inside of the bankfull channel (see Chapter 6). Central and transverse bars are common, and related to the high sediment supply from streambanks and the high width/depth ratios. Stream bank erosion rates are very high due to side slope rejuvenation and mass-wasting processes which enhance the fluvial entrainment. Riparian vegetation plays a marginal role in streambank stability due to the typically very high bank heights, which extend beyond the rooting depth of riparian plants. Exceptions to this are the F4 stream types in the Northeast, Northwest, and Southeast United States where the relatively longer growing seasons and ample precipitation results in the establishment of riparian vegetation that tends to cover the entire slope face of channel banks.

THE MORPHOLOGICAL DESCRIPTION

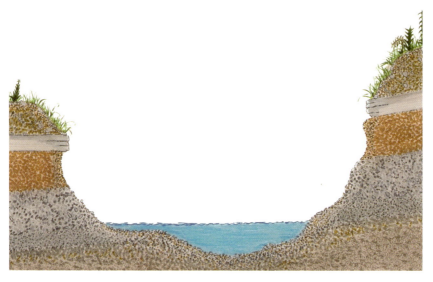

DELINEATIVE CRITERIA (F4)

Landform/soils: The F4 stream type is associated with deeply entrenched, structurally controlled, gentle gradient valleys and gorges. The F4 stream type is associated with highly weathered bedrock or depositional soils involving a combination of river downcutting and uplift of valley walls.

Channel materials: Gravel dominated channel with some cobble and sand accumulations. Streambanks are generally eroding unless stabilized with massive riparian vegetation.

Slope Range: < .02   **Entrenchment Ratio:** < 1.4

Width/depth Ratio: > 12   **Sinuosity:** > 1.2

THE MORPHOLOGICAL DESCRIPTION

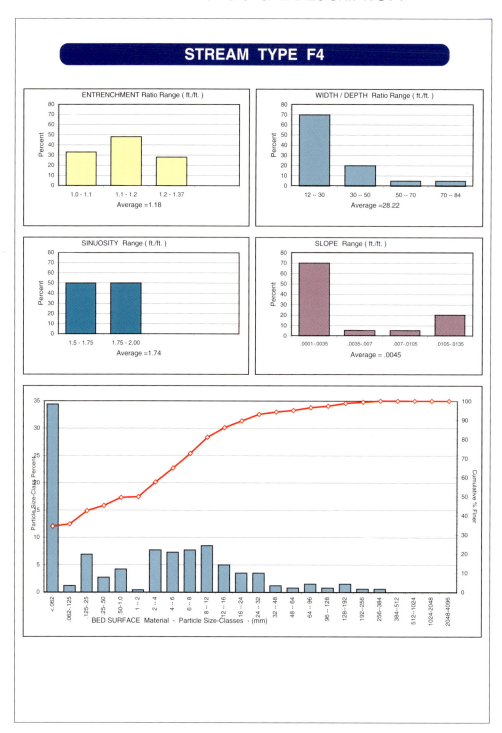

THE MORPHOLOGICAL DESCRIPTION

F4 - Colorado

F4 - California

F4 - Texas

MORPHOLOGICAL DESCRIPTION AND EXAMPLES OF STREAM TYPES

F5 Stream Type

The F5 stream type is a sand dominated, entrenched, meandering channel, deeply incised in gentle terrain. The "top of banks" elevation for this stream type is much greater than the bankfull stage which is indicative of the deep entrenchment. The F5 stream type can be deeply incised in alluvial valleys or in lacustrine deposits, resulting in the abandonment of former floodplains. The F5 stream channels are found in valley types IV, VI, VIII, X, and XI. The F5 channels have slopes that are generally less than 2%, exhibit riffle/pool bed features, and have width/depth ratios that are high to very high. The dominant channel materials are sand with lesser accumulations of gravel and some silt-clay. Sediment supply in the F5 stream types is moderate to high, depending on stream bank erodibility conditions. Depositional features are common in this stream type, and over time, tend to promote development of a flood plain inside of the bankfull channel (see Chapter 6). Central and transverse bars are common, and related to the high sediment supply from streambanks and the high width/depth ratios. Stream bank erosion rates are very high due to side slope rejuvenation and mass-wasting processes which enhance the fluvial entrainment of eroded bank materials. Riparian vegetation plays a marginal role in streambank stability due to the typically very high bank heights which extend beyond the rooting depth of riparian plants. Exceptions to this are the F5 stream types in the Northeast, Northwest, and Southeast United States where the relatively longer growing seasons and ample precipitation results in the establishment of riparian vegetation that tends to cover the entire slope face of channel banks.

THE MORPHOLOGICAL DESCRIPTION

DELINEATIVE CRITERIA (F5)

Landform/soils: The F5 stream type is associated with deeply entrenched channels in alluvium or in structurally controlled, gentle gradient valleys and gorges. The F5 stream type is associated with highly weathered rock or depositional soils involving a combination of river downcutting and/or uplift of the valley walls.

Channel materials: This is a sand dominated channel, both bed and streambanks.

Slope Range: < .02   **Entrenchment Ratio:** < 1.4

Width/depth Ratio: > 12   **Sinuosity:** > 1.2

THE MORPHOLOGICAL DESCRIPTION

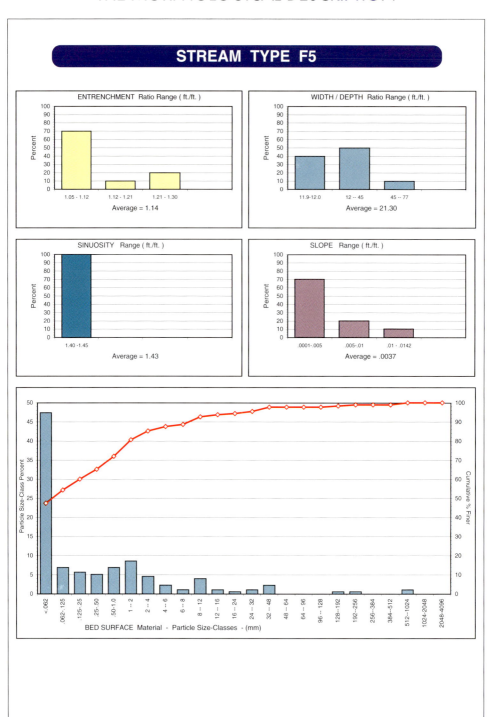

THE MORPHOLOGICAL DESCRIPTION

F5 - Colorado

F5 - Texas

F5 - Maryland

165

MORPHOLOGICAL DESCRIPTION AND EXAMPLES OF STREAM TYPES

F6 Stream Type

Texas

The F6 stream types are entrenched, meandering, gentle gradient streams deeply incised in cohesive sediments of silt and clay. The F6 stream channels have very high width/depth ratios, moderate sinuosities, and low to moderate meander width ratios. The "top of bank" elevation for this stream type is much greater than the bankfull stage which is indicative of the deep entrenchment. The related landforms are often seen as terrace bounded alluvial valleys, deltas, and coastal plains. The F6 stream channels are found in valley types IV, VIII, X and XI. Depositional soils in these valleys often originate from fine alluvial, eolian (loess), and lacustrine deposits. The F6 stream banks are relatively more stable than the F3, F4, or F5 stream banks, due to their inherent cohesive nature and ability to "stand" much steeper. Deep rooted riparian vegetation is much more effective at maintaining stability in the cohesive bank materials. However, mass wasting due to bank saturation/liquefaction and collapse is still a prevalent process in hydro-physiographic provinces where the composition of riparian vegetation is poor and natural densities have been reduced. The F6 stream systems produce relatively low bedload sediment yields due to the lack of coarse material in the channels, thus, excessive bar deposition is not generally observed with the F6 stream type. These stream types are very sensitive to disturbance and adjust rapidly to changes in flow regime and sediment supply from the watershed.

THE MORPHOLOGICAL DESCRIPTION

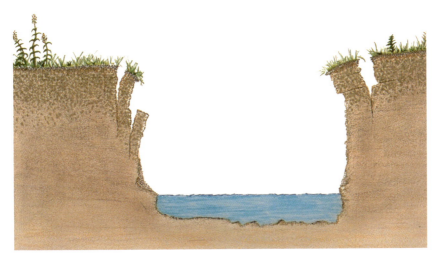

DELINEATIVE CRITERIA (F6)

Landform/soils: The F6 stream type is associated with deeply entrenched channels in alluvium or in structurally controlled, gentle gradient valleys and gorges. The F6 stream type is associated with highly weathered rock or depositional soils involving a combination of river downcutting and/or uplift of the valley walls. Cohesive soils with occasional mass-wasting slump blocks.

Channel materials: Silt and or clay

Slope Range: < .02

Entrenchment Ratio: < 1.4

Width/depth Ratio: > 12

Sinuosity: > 1.2

THE MORPHOLOGICAL DESCRIPTION

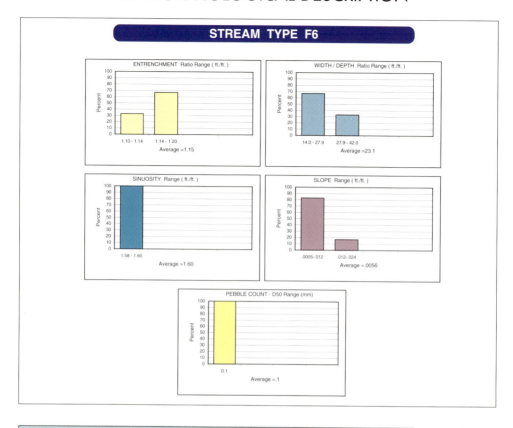

F6 - Colorado

THE MORPHOLOGICAL DESCRIPTION

F6 - Virginia

F6 - Texas

F6 - Maryland

MORPHOLOGICAL DESCRIPTION AND EXAMPLES OF STREAM TYPES

G1 Stream Type

G1 stream channels are deeply entrenched into bedrock and have moderate channel gradients, low width/depth ratios, and randomly spaced steps and plunge pools. The G1 stream type patterns, profiles, and dimensions are structurally controlled and related to the presence of faults and joints or erosion into highly weathered bedrock. The stream type is very stable, with limited rates of lateral or vertical adjustment. The G1 stream type is a step/pool system with low sediment storage capacities and a low sediment supply, due to the stable nature of the channel bed and bank materials. The G1 is similar to the A1 stream type with the exception that the G1 occurs primarily on moderate slopes and has a slightly higher channel sinuosity. G1 channels can also occur as narrow, deep gorges on larger rivers, where the reach has a gradient of 2-4 percent and produces the more difficult class 4 and 5 rapids, which are often used for recreational boating.

THE MORPHOLOGICAL DESCRIPTION

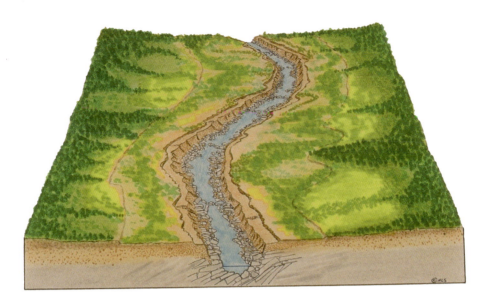

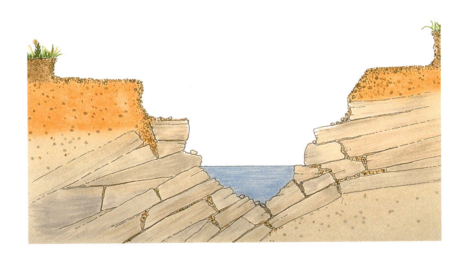

DELINEATIVE CRITERIA (G1)

Landform/soils: The G1 stream type is associated with moderately steep, structural controlled, narrow valleys. They are similar to the A1, but not as steep.

Channel materials: Bedrock with a mixture of boulders and cobble with some minor amounts of gravel.

Slope Range: < .04

Entrenchment Ratio: < 1.4

Width/depth Ratio: < 12

Sinuosity: > 1.2

171

THE MORPHOLOGICAL DESCRIPTION

G1 - Arizona

G1 - Colorado

G1 - Idaho

THE MORPHOLOGICAL DESCRIPTION

G1 - Colorado

G1 - Oregon

G1 - New Mexico

MORPHOLOGICAL DESCRIPTION AND EXAMPLES OF STREAM TYPES

G2 Stream Type

Colorado

The G2 stream channels are deeply entrenched, slightly meandering, step/pool systems, with the dominant channel bank and bed materials appearing as boulders. The G2 is very stable, with moderate channel gradients of 2 to 4 per cent and a low width/depth ratio. The "slope continuum" concept is applied to the stream type description if the observed reach exhibits slopes of less than 2 percent. Such a reach would be designated as a G2c, while maintaining a similar morphology, dimension, and pattern. For those channels with flatter slopes, the width/depth ratio and sinuosity tends to slightly increase above values observed for G2 channels with steeper slopes. The G2 stream type exhibits a channel bed dominated by boulder materials, while the channel banks generally have a higher percentage of cobble, gravel, and some sands mixed with scattered boulders. The G2 stream type is associated with very coarse alluvial fans, boulder debris from landslides, wedges below talus fields, colluvial deposits from up-slope gravitational erosion, and structurally controlled slopes. The G2 stream types can also occur as narrow, deep gorges on larger rivers, where the reach has a gradient of 2-4 percent, and produces the more difficult class 4 and 5 rapids which are often used for recreational boating.

THE MORPHOLOGICAL DESCRIPTION

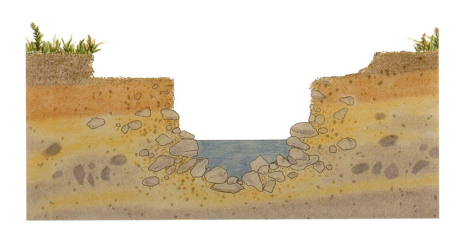

DELINEATIVE CRITERIA (G2)

Landform/soils: The G2 stream type is associated with structural controlled, narrow, moderately steep valleys. They are similar to the A2 stream types but not as steep.

Channel materials: Boulder dominated channel with cobble and gravel.

Slope Range: < .04

Width/depth Ratio: < 12

Entrenchment Ratio: < 1.4

Sinuosity: > 1.2

THE MORPHOLOGICAL DESCRIPTION

G2 - Colorado

G2 - Arizona

THE MORPHOLOGICAL DESCRIPTION

G2 - Colorado

G2 - Colorado

MORPHOLOGICAL DESCRIPTION AND EXAMPLES OF STREAM TYPES

G3 Stream Type

The G3 stream type is deeply incised in depositional material primarily comprised of an unconsolidated, heterogenous mixture of cobble, gravel, and sand. The G3 stream type is highly unstable due to the very high sediment supply available from both upslope and channel derived sources. The G3 channels have a moderate gradient, a low width/depth ratio, a characteristic step/pool morphology, and low sinuosities except when deeply incised in a previously sinuous channel. Bank erosion and bedload transport is typically very high in the G3 stream channels due to the combined effects of low width/depth ratios, moderate channel gradients, and the high sediment supply. The ratio of bedload to total sediment load often exceeds 50%. The observed effects of vertical and lateral instability processes are primarily due to the combination of high streamflow energy and high available sediment supply. The G3 stream types are usually found in landform features such as alluvial fans, and landslide debris, and often seen as headcut gullies deeply incised in meadows, fluvial terraces, and in the bottom of previous channels. These stream types are very sensitive to disturbance and tend to make significant adverse channel adjustments to changes in flow regime and sediment supply from the watershed.

THE MORPHOLOGICAL DESCRIPTION

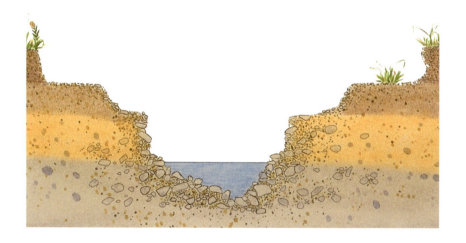

DELINEATIVE CRITERIA (G3)

Landform/soils: The G3 stream type is associated with moderately steep, fluvial dissected landforms, alluvial fans or down cut in alluvial or colluvial valleys. Soils are a heterogeneous mixture of unconsolidated non-cohesive material generally in alluvium and colluvium.

Channel materials: Cobble dominated channel with a mixture of gravel and sand.

Slope Range: < .04   **Entrenchment Ratio:** < 1.4

Width/depth Ratio: < 12   **Sinuosity:** > 1.2

THE MORPHOLOGICAL DESCRIPTION

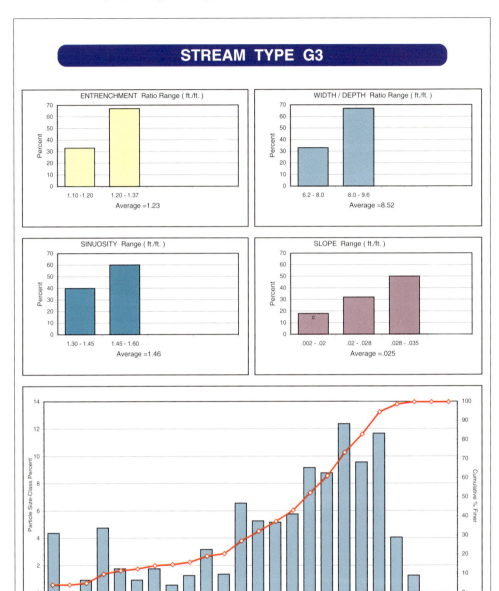

THE MORPHOLOGICAL DESCRIPTION

G3 - Colorado

G3 - California

MORPHOLOGICAL DESCRIPTION AND EXAMPLES OF STREAM TYPES

G4 Stream Type

Colorado

The G4 stream type is deeply incised in depositional material primarily comprised of an unconsolidated, heterogenous mixture of gravel, some small cobble, and sand. The G4 stream type is very unstable due to the very high sediment supply available from both upslope and channel derived sources. The G4 stream channels have a moderate gradient, a low width/depth ratio, a characteristic step/pool morphology and low sinuosities; except when deeply incised in a previously sinuous channel. Pools are often filling with bedload, as the potential for sediment storage is high. Bank erosion and bedload transport rates are typically high in the G4 stream channel due to the combined effects of low width/depth ratios, moderate channel gradients, and the high sediment supply. The ratio of bedload to total sediment load often exceeds 50%. The observed effects of vertical and lateral instability processes are primarily due to the combination of high streamflow energy and high available sediment supply. The G4 stream types are usually observed in landform features such as alluvial fans, landslide debris, and are often seen as deeply incised headcut gullies in meadows, fluvial terraces, and in the bottom of previous channels. These stream types are very sensitive to disturbance and tend to make significant adverse channel adjustments to changes in flow regime and sediment supply from the watershed .

THE MORPHOLOGICAL DESCRIPTION

DELINEATIVE CRITERIA (G4)

Landform/soils: The G4 stream type is associated with moderately steep, fluvial dissected landforms, alluvial fans or down cut in alluvial or colluvial valleys. Soils are a heterogeneous mixture of unconsolidated non-cohesive material generally in alluvium and colluvium.

Channel materials: Gravel dominated channel with mixtures of sand and some cobble.

Slope Range: < .04

Entrenchment Ratio: < 1.4

Width/depth Ratio: < 12

Sinuosity: > 1.2

THE MORPHOLOGICAL DESCRIPTION

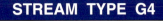

STREAM TYPE G4

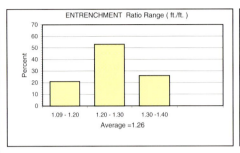

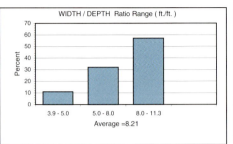

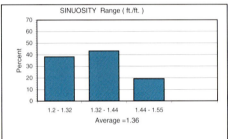

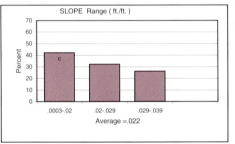

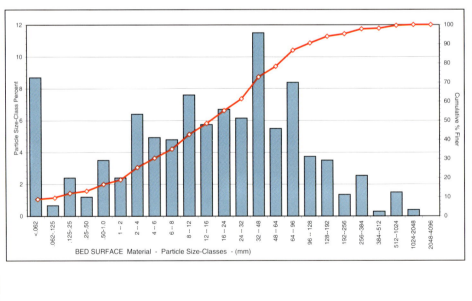

THE MORPHOLOGICAL DESCRIPTION

G4 - Nevada

G4 - California

G4 - Maryland

MORPHOLOGICAL DESCRIPTION AND EXAMPLES OF STREAM TYPES

G5 Stream Type

New Mexico

The G5 stream type is an entrenched, moderately steep, step/pool channel deeply incised in sandy materials. Channel sinuosities are relatively low, as are width/depth ratios. These "sandy gully" stream types transport great amounts of sediment due to the ease of particle detachment and fluvial entrainment. The G5 stream channels are generally in a degradation mode derived from near continuous channel adjustments, due to excessive bank erosion. Bedload transport rates can easily exceed 50 % of total load; with active, extensive, consistent channel erosion more typical than not. Exceptions may occur where very dense woody vegetation helps stabilize the toe of the stream bank slopes. The G5 stream type is similar in character to A5 channels, except G5 channel gradients are less than 4% and, tend to be more sinuous with somewhat higher width/depth ratios, due to the gentler channel slopes. The "slope continuum" concept is applied for the "gully" stream types if the observed reach exhibits slopes less than 2 %. Such a reach is given the designation of G5c. The lower gradient gully reaches are generally observed developing within a previously meandering, low gradient system with floodplains such as a C5 situated in wide alluvial valleys. These stream types are very sensitive to disturbance and tend to make significant adverse channel adjustments to changes in flow regime and sediment supply from the watershed.

THE MORPHOLOGICAL DESCRIPTION

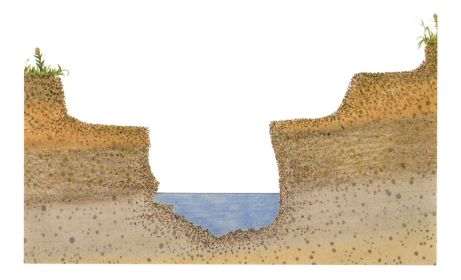

DELINEATIVE CRITERIA (G5)

Landform/soils: The G5 stream type is associated with moderately steep, fluvial dissected landforms, alluvial fans or down cut in alluvial or colluvial valleys. Soils are a heterogeneous mixture of unconsolidated non-cohesive material generally in alluvium, and colluvium, eolian (sand) deposition and residual soils such as those derived from grussic granite.

Channel materials: Sand dominated channel with mixtures of gravel and some silt/clay.

Slope Range: < .04

Entrenchment Ratio: < 1.4

Width/depth Ratio: < 12

Sinuosity: > 1.2

THE MORPHOLOGICAL DESCRIPTION

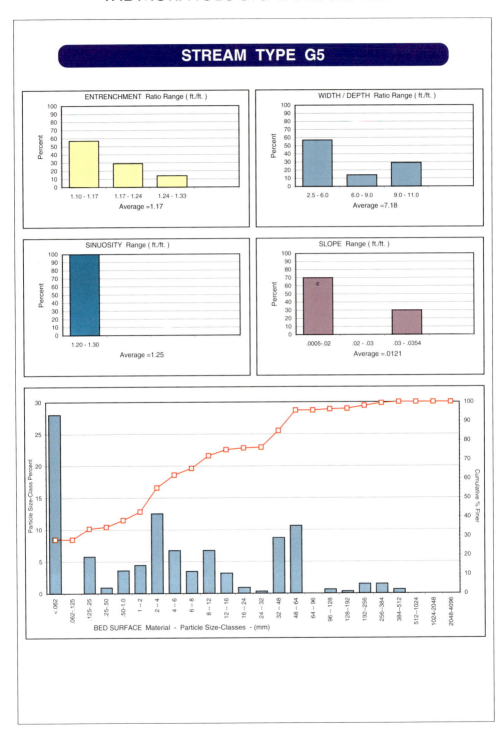

THE MORPHOLOGICAL DESCRIPTION

G5 - Colorado

G5 - Arizona

G5 - Nevada

MORPHOLOGICAL DESCRIPTION AND EXAMPLES OF STREAM TYPES

G6 Stream Type

The G6 stream type is an entrenched gully system with gentle to moderately steep channel gradients; that is deeply incised in cohesive materials of silts and clays. Bedload sediment transport rates are relatively low, and replaced by high washload and suspended sediment yields that commonly occur within the stream type. The bed features are generally observed as an unstable, degrading step/pool morphology. The dominate lithology for the G6 types include shales and depositional environments such as fans, deltas, lacustrine landforms, and other features that have cohesive, silt/clay deposits. Streambank erosion processes acting on the typically steep banks produce very high amounts of erodible material, especially within delta and lacustrine landforms. Woody riparian vegetation can have a bank stabilizing tendency if the vegetation densities are very high. The G6 stream types are very sensitive to disturbance and tend to make significant adverse channel adjustments to changes in flow regime and sediment supply from the watershed. The G6 stream type is generally considered to be experiencing near continuous degradational processes. It is not unusual to observe channel gradients of less than 2% (G5c), or even channel slopes less than .1% (G5c-).

THE MORPHOLOGICAL DESCRIPTION

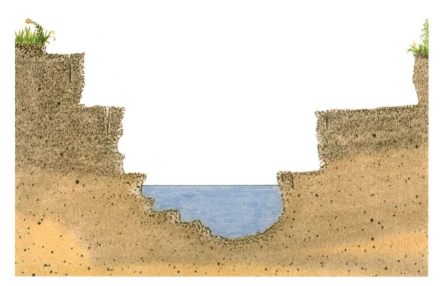

DELINEATIVE CRITERIA (G6)

Landform/soils: The G6 stream type is associated with moderately steep, fluvial dissected landforms, alluvial fans or down cut in alluvial or colluvial valleys. Soils are cohesive materials generally in alluvium, colluvium, eolian deposits (loess), and residual soils such as those derived from shales.

Channel materials: Silt/clay dominated channel with mixtures of gravel and some silt/clay.

Slope Range: < .04 **Entrenchment Ratio:** < 1.4

Width/depth Ratio: < 12 **Sinuosity:** > 1.2

THE MORPHOLOGICAL DESCRIPTION

G6 - Colorado

G6 - Utah

G6 - California

THE MORPHOLOGICAL DESCRIPTION

G6 - Utah

G6 - Colorado

G6 - Nevada

TO USE THIS FORM:
Please make photocopy and fill out, then mail or fax to the location below.

To Order:

Wildland Hydrology Books
1481 Stevens Lake Rd.
Pagosa Springs, CO 81147
Phone: (970) 731-6100 or 6101
Fax: (970) 731-6105

Company _____

Name _____

Street _____

City _____ State _____ Zip _____

Phone _____ Fax _____

Payment: ❏ MC   ❏ Visa   ❏ AMEX   ❏ Check

Card# _____

Exp. Date: _____ / _____ / _____

Issuing Bank: _____

Signature _____

TITLES	COST PER BOOK	NUMBER OF BOOKS	COST
Applied River Morphology S&H – $7.50/1book - $4.50/additional books	$89.95	_____	$_____
The Reference Reach Field Book S&H – $4.00/1book - $2.50/additional books	$19.95	_____	$_____
Field Guide To Stream Classification S&H – $6.00/1book - $3.50/additional books	$55.00	_____	$_____
TOTAL		_____	$_____

NOTE: Orders will be shipped UPS. Higher shipping rates outside of the USA.
NO PURCHASE ORDERS ACCEPTED